The Young Farmer: Some Things He Should Know

by Thomas Forsyth Hunt

CONTENTS

CHAPTER

THE YOUNG FARMER: SOME THINGS HE SHOULD KNOW

CHAPTER I

ESSENTIALS OF SUCCESS

Columella, the much traveled Spanish-Roman writer of the first century A. D., said that for successful farming three things are essential: knowledge, capital and love for the calling. This statement is just as true today as it was when written 1900 years ago by this early writer on European agriculture.

Every man who loves the calling and has an ambition to become a successful farmer should understand that no two of these essentials are sufficient, but that all three are necessary. Although this is so simple as to be almost axiomatic, it is indeed surprising how few people believe a knowledge of farming is really essential to success.

America is strewn with cases of failure, in farming, by men investing capital acquired in other business. In nine cases out of ten failure has been due to lack of knowledge of farming.

There is known to the writer an expert mineralogist and metallurgist. On the subject of coal and gold mining he can give the most valuable information. His advice is constantly sought on all such matters. Instead of investing his money in mining, on which he is a recognized authority, he has invested it in a farm, about which he knows next to nothing. He has not even had the advantage of being raised on a farm, since his father was a railroad man.

A mechanical engineer remarked that if he had $25,000 he would invest it in a farm. This man is supposed to be an expert in business methods as applied to manufacturing in general, and he is especially conversant with the manufacture and trade in automobiles. About all he has seen of farming he has observed from the window of a Pullman car or from the steering wheel of an automobile. Instead of investing his earnings in some manufacturing business, about which he has spent years of study and in which he has had some training, he would invest it in farming, of which he has only the most rudimentary knowledge, if only he had sufficient capital. As a matter of fact, he is more in need of knowledge than of capital.

Even farmers of experience do not always realize the training required to succeed in farming. A letter was received by the dean of a certain agricultural college saying that a graduate of another agricultural college had taken one of the poorest farms in his neighborhood and was raising better potatoes than anyone else could raise. The letter asked that information be sent by return mail as to how this young man could be beaten in raising potatoes. Of course the answer had to be sent that while information upon raising potatoes could easily be supplied, although not in the limits of an ordinary letter, the training in observation, judgment and reasoning faculties essential to meet the daily problems as they arise could not be supplied.

There is no objection to men of other vocations adopting farming as an avocation if they can afford it. It is a rational form of pleasure for wealthy people, and one in which they can often be of great service. This cannot be said of all forms of relaxation. Wealthy men have been of special service to the cause of agriculture by promoting the breeding of improved live stock. Men in other callings should clearly understand, however, that if they have a farm merely as a place to spend a week end, that they may expect to find the financial returns unsatisfactory.

To no one is there more significance in the old school aphorism "knowledge is power" than to the young man who is to become a farmer. While it is not necessary to be educated in schools in order to gain knowledge, yet the schoolroom with all its limitations is usually the most economical and most efficient method of acquiring certain forms of knowledge essential to every successful man or woman. A farm-to-farm canvass of a certain region of the state of New York discloses the fact that farmers with college training are obtaining a higher income from their farms than those whose school days ended with high school. Similarly, those who have finished the high school are more prosperous financially than those who never advanced beyond the grades. The investigation showed, for example, that with the farmers under observation the high school education was equivalent to $6,000 worth of 5% bonds. Farming is an occupation requiring keen observation, sound judgment and accurate reasoning, all attributes which are strengthened greatly by proper education. This is so true that many men, perhaps most men, are forty before they have grasped the problems which the truly successful farmer must solve.

A considerable part of the knowledge essential to success in any pursuit is acquired by actually working at the occupation, or, as we say, by practical experience. Some features of any occupation can be obtained in no other way. A preliminary education may, however, greatly reduce the time necessary to acquire even this practical experience. For example, a course in shop work as taught in technical high schools and colleges, requiring two hours a day for five months, may shorten the time of apprenticeship by one or more years, in acquiring the trade of carpenter or iron worker. In the same manner a course in butter making, cheese making or floriculture, may shorten the time required to obtain the necessary practical details by ten months or even more. Eventually, also, the man thus trained will be the better man.

If the industrial activities of the world be divided into farming, mining, manufacturing, trade and transportation, it will be noted at once that farming is the only one which deals with living things. In fact, the definition of agriculture, in its broadest sense, is the economic production of living things. The farmer is thus brought face to face with some of the most difficult and intricate problems with which the human race has to grapple. It is this fact that makes farming, in some ways, the most uncertain as well as the most fascinating occupation known to man. The fact that the farmer is dealing with living things puts his occupation in a class by itself for a number of reasons, one of which is germane to the subject of this chapter.

In most occupations a larger part of the knowledge necessary to success can be acquired by doing than is the case in farming. Locomotive engineers are trained for their responsible duty while firing the engine. The brakeman becomes a conductor by assisting the latter. A bank cashier is usually a promoted bank clerk. Each obtained the knowledge essential to success largely by oft-repeated performance.

While, of course, there is much the farmer can learn only by experience, there are many things essential to his success that the mere performance of the necessary farm operations will not teach him. Spreading manure will never teach him that stable manure should be supplemented with phosphoric acid in order to get the best results. The growing of clover will not teach him that mineral fertilizer may keep up the fertility of the soil where clover grows luxuriantly and occurs in the rotation at definite intervals. Feeding cattle will not teach him that a good ration for milch cows is one containing one pound of digestible protein to seven pounds of digestible carbohydrates, provided it is palatable and, at least, two-thirds of the total ration is digestible. Nor will the feeding of such a ration teach the farmer how to calculate the most economical ration from feeding stuffs at current prices. The cause of potato blight and the methods of combating it cannot be learned from the operation of planting and cultivating potatoes.

These are only a few illustrations--they might be multiplied indefinitely--to show that farming is peculiar in that performance of the daily duties does not give the knowledge essential to success in the same measure that it does in such occupations as banking, trade and transportation. Yet, curiously enough, while no man would undertake to run a locomotive engine or perform the duties of cashier of a bank without thorough training, there are many who will undertake to farm without education or knowledge of the business.

The young man who intends to become a farmer should fully understand that if farming is not a business worthy of a thoroughly educated man, it is not a business worthy of him; because every young man is worthy of a thorough education, provided he is a man of clean habits and good purposes. Do not allow yourself to be persuaded that you lack ability to acquire a good education. All you require is opportunity, determination and honesty of intention.

Farming is worthy, moreover, of the most highly educated as well as the most capable. If lack of means prevents a young man from taking a four-years' training in agriculture, he will find a two years' course offered by many of the state agricultural schools. While it is obviously impossible to give in two years as much training as in four years, these two years' courses contain the more technical subjects and are usually very thorough and efficient. No young man, no matter how thorough his previous training, need hesitate to pursue one of them.

There are, however, young men who cannot spare the time and expense of even two years' training. For such many state agricultural colleges offer winter terms of eight to twelve weeks. These courses are arranged to allow the student to specialize along some particular line. The better prepared the man is who enters these winter courses the more he will benefit by them. This leads to the caution that such courses should not be substituted for the education offered in the public schools, but should only be sought after all the opportunities for education at home have been exhausted.

For the somewhat older young man who is now farming and cannot leave his farm or for the younger man as a preparation for the short courses, one or more correspondence courses will be found useful. Not all colleges conduct correspondence courses, but fortunately those who do will accept students from other states on equal terms. There are many persons who will testify to their helpfulness.

Every young farmer should have a carefully selected library of standard books on agriculture, not only for reading but for reference. An instance of the value of a standard book of reference came recently to the attention of the writer. An educated young farmer in Iowa paid $2.50 for a peck of crimson clover seed which he sowed in the spring in his oats. A reference to any standard publication on forage crops costing less than the peck of seed would have disclosed to him the probable hopelessness of success under the conditions named.

The books to include as well as to exclude from a select list will depend upon the previous training of the man making the purchase, the character of the farming to be pursued, and, to some extent, to the section of the country where the farm is located. Any bookseller can secure catalogs issued by firms making a specialty of publishing agricultural books. For the average reader these catalogs are sufficient to enable one to make intelligent purchases.

Every farmer should take one or more agricultural journals. At present journals are published on every phase of agriculture and many of them are of high character. Publishers are always glad to send sample copies free of charge. By examining these copies intelligent selection may be made.

The writer of this book has had rather unusual opportunity during more than a quarter of a century of observing the influence of education upon the success, financial and otherwise, of those who engage in farming. As the result of these observations he wishes to urge every young man to allow no one to persuade him that because he is to be a farmer, he does not need a thorough education. Remember that you have but one life to live, and if you let the golden opportunity pass, the mistake can never be rectified. No man ever regretted that he had too much education--thousands have regretted the lack of it.

Every young man, no matter what his occupation is to be, should receive some school training, however little it may be, every year until he reaches the age of majority. Otherwise the age of majority should be changed. In no occupation is this more important than in farming, because the operations involved in farming fail to develop certain attributes necessary to the largest success.

A man cannot have a mind too well trained, although it is possible that he may have too much undigested information. The mental condition may not be unlike the physical condition of the man who is burdened with too many clothes. When in action he may need to strip his mind of unnecessary information in order to make the most efficient mental effort.

CHAPTER II

MEANS OF ACQUIRING LAND

Of the three essentials to successful farming--capital, knowledge and love for the calling--only the first can be obtained on credit, and this only in part. Usually when a man desires to buy a farm he must have, at least, one-third of his desired investment in cash. The amount to be invested will include, not only the cost of the land, but the cost of the necessary equipment of the farm. The percentage of the total capital which may be borrowed, however, will depend on many circumstances and is usually a matter of first importance. No man should borrow more than a banker or other reputable business man considers a safe investment.

Usually there is no better counselor as to a safe investment than the local banker. The banker should, and generally does, stand in much the same relation to the financial welfare of the community as the physician to its physical, the minister to its moral and spiritual welfare. The inexperienced person, even if he does not need to borrow money, would do well to consult some responsible banker in the neighborhood before making an investment in farm lands.

The young man should, as early as possible in life, open an account with the local bank, not merely for the sake of the habit of saving which this will encourage, but in order to come into personal business relations with the banker. Instead of concealing from the bank his business operations, he should seek the advice of his banker on all important financial matters.

On an average, every farm changes hands at least three times in a century. Every farm, therefore, must be acquired by purchase, inheritance or gift at more or less irregular intervals. In the neighborhood in which the author was born, there is not a farm but has changed hands since he can remember. In many cases the farm is now in the possession of a son; in some instances in that of a grandson of the owner as known by the writer in his boyhood days. In this particular community the acquirement of a farm by a person not related to the former owner has occurred in relatively few instances.

As a rule, when the farm has been acquired by a son, the latter has operated the farm as tenant or partner for a period previous to his ownership and during lifetime of the father. In some instances the son has boarded with the parents or the parents with the son and his wife; or, in the case of a daughter, with the daughter and son-in-law.

Where there are several heirs, as is apt to be the case, the son operating the farm is required to purchase or rent the interest of the other heirs, unless the farm is large enough to be divided, which is less seldom the case than is popularly supposed. Thus, if there are 200 acres of land worth $50 an acre, and five heirs, the young farmer may inherit $2,000, and be required to assume the remaining $8,000 as an obligation. He may borrow this money at the bank, placing a mortgage upon the farm, thus settling with the other heirs at once. Or he may pay the other heirs rent on their share of the farm. In any case he will, if successful, gradually cancel his obligation and become owner of the farm. That no heir is willing to assume this responsibility is the most common reason for a farm changing from one family to another, and the disruption of community interests.

The customary, or normal, method of acquiring land has been and still is a combination of tenancy, inheritance and mortgage. Without some tenant system and without the farm mortgage, it would be impossible for the average young man to acquire a farm. That men are constantly advancing from farm tenant to landowner is shown by statistics giving the percentage of tenants by ages. The majority of farmers under 30 are renters. Most farmers over 45 are owners of farm land. Thus in Illinois, in 1900, approximately 75% of the farmers under 25 years of age rented their farms, while less than 20% of the farmers over 55 years of age were tenants.

The question for the young man to consider is not what effect the tenant system has upon the welfare of the nation or what political ills may be connected with farm mortgages, but how to make use of these necessary and beneficent agencies for the acquirement of a farm. A system of tenancy which leads to absent landlordism and a permanent tenant class is thoroughly vicious, while a practice which enables a man to become, within a reasonable period, a land-owning farmer is a thoroughly approvable and, indeed, necessary method of acquiring land.

As already indicated, most young men will need in some form or other to employ more capital than they possess when they start farming. They must, therefore, determine what is the best form of obtaning the necessary capital, viz.: whether to borrow the money on a farm mortgage, or whether to use the capital someone else has invested in a farm by paying him rent for it. The conditions of tenancy in this country are often not the most fortunate, yet the young man of character may well find, for a time, at least, it would be best for him to rent a farm and invest his own capital in the necessary machinery and live stock to conduct it properly.

Much will depend on the character of the arrangement which may be made. Usually more favorable terms can be secured from landlords owning large numbers of farms than from the owner of one or two farms. The large landowner is content with a moderate income from each farm, because in the aggregate his income is sufficient for his needs, while the retired farmer who must live off the proceeds of a single farm is apt to drive a hard bargain and may not be over particular concerning the maintenance of said farm. The writer knows a farmer who owns a good farm purchased from the proceeds of a rented farm. He continues to live on the rented farm and rents his own, because, it is said, his landlord is willing to make him more favorable terms than he makes to his tenant.

The more capable the tenant the more favorable the terms he may exact. Certain tenants are in demand and can have their choice of farms. A prosperous-looking man was pointed out recently as an example of a tenant capable of buying a farm in one of the most highly developed counties in the United States. It was stated that as a renter he could have his choice of any farm in the county, but that he did not have a dollar invested in farm land. Possibly he invests his surplus earnings in stocks and bonds.

It is not the present purpose to determine the relative merits of the different systems of land tenure, but to try to be helpful to the beginners by discussing the usual practices in order that he may know whether the arrangement he is considering is customary and whether it is likely to prove satisfactory.

Every third farm in the United States is rented under one of three methods:

1. A definite money rent may be paid, ranging from $2 to $6 an acre for land on which the ordinary, staple crops are raised. Perhaps $3 to $4 is more commonly paid for such land.

2. In the South it is common for the landlord to require a definite number of

pounds of cotton per acre or a certain number of bales of cotton for a one or two-mule farm, as the case may be. This is classified by the census authorities as "cash rent," but will here be called "crop rent." Crop rent is less common than either cash or share rent in the northern and western states, although perhaps the most common form in the South. Crop rent, however, is met with in some sections, as in western New York where certain large landowners require a definite number of bushels of wheat, oats or maize and make certain stipulations as to hay and straw. They charge a cash rent for pasture.

3. Much the most common form of tenancy, however, is that where a certain percentage or share of the product is given the landlord for the use of the land.

Before entering into a discussion of the customary conditions under which land is rented on shares it may be helpful to point out the fundamental differences between cash rent, crop rent and share rent. In case of cash rent, the landlord takes no risk, either as to the price or the amount of product. In the case of crop rent, he shares the risk as to the variation in price, but not as to the amount of crop raised. The latter may depend upon the clemency of the weather or upon the industry and skill of the tenant. In the case of share rent, both landlord and tenant share equally as to variation in the price and the amount of product.

Three forms of share rent may be recognized:

(a) Where landlord furnishes only real estate (land and buildings), the tenant supplying everything else, including teams, machinery, labor, seeds and fertilizers. Under these conditions it is customary for the landlord to receive one-third and the tenant two-thirds of the crop raised or the product produced.

(b) The second form of share rent is where the landlord furnishes the real estate; the tenant supplies teams, tools and labor, while the landlord and tenant own equally all live stock other than teams, and bear equally all other expenses, as for seeds, fertilizers and cost of threshing. Under this system, it is customary for landlord and tenant each to receive one-half of all sales. As each owns one-half of all the live stock (teams excepted), each shares equally in all increase. The landlord pays for the cost of permanent improvements such as new buildings, fences, repairs and drainage. The tenant, in making these improvements, in some cases, agrees to furnish two days' labor for one day's pay. The theory is that, while the increased value of the real estate is of advantage only to the landlord, the improved facilities are of some benefit to the tenant. Since he can do this work at odd times when not otherwise employed, he can afford to take a generous view of the matter. It is obvious that if he remains on the farm long enough the tenant will come into his share of the benefit, while if he intends to leave the farm soon he may not. There is in the mind of the writer a prosperous tenant who, after eighteen years on a single farm, declared he had no desire to make a change, and doubtless there are thousands of similar instances.

Under the plan in which the tenant furnishes everything except the real estate, the tendency of the farm is apt to be downward both as to the improvements and the crop-producing power of the soil. The interests of the landlord and tenant are not mutual. This condition of tenancy leads to growing only those crops which can be readily sold from the farm and to frequent changes of the tenant, with its accompanying auction sales of property. In one region, where this system prevails, it has been facetiously remarked that each tenant has a sale every year to determine how much he is worth. It is less trouble than taking an inventory.

In the second form of share rent, the interests of landlord and tenant are more nearly mutual. Under this system, animal husbandry is possible, which, generally, involves pasturing and feeding a considerable part of the crops upon the farm, and even the purchase of nitrogenous by-products. All this leads to permanency of tenant, since the landlord and tenant are both interested in the live stock and other personal property, which cannot be divided, with economy, each year. It is interesting to note that the house is the least likely to be kept in repair. The improvement of the barns and fences or the laying of tile drains increases the landlord's income, but he has no financial interest in the house, so long as the tenant is willing to live in it.

There are, of course, many variations in the arrangement of details between the landlord and tenant. On many dairy farms in the northeastern states it is customary for the landlord to own the cows. While the landlord and tenant share equally from the sale of milk, butter or cheese, in such cases the increase in the herd belongs to the owner of the land. Hence, money from the sale of any animal, old or young, goes to him. This is because the landlord must keep up the herd. If a cow is sold, he must furnish another to take her place.

(c) The third type of tenant farming is where the tenant furnishes nothing but his labor and managerial ability, and receives a share of the sales, which may be one-third. This is rather an unusual type of tenancy, since, where the landlord furnishes all the capital, it is much more common to employ a farm manager at a monthly wage. The wage varies greatly, but is seldom below forty dollars or above seventy-five dollars per month without board, especially to those who have not hitherto had much managerial experience.

Various attempts at profit sharing have been made. A recent instance is of a young married man taking 160 acres of tillable land where the landlord has a fairly well-stocked farm. The young man is to have a house and everything in the way of living the farm can furnish. He is to receive $20 a month and one-half the net proceeds, or, what is called in Chapter XI, the farm income. In considering a contract of this kind it is necessary to make a careful distinction between: (1) Gross sales, (2) net proceeds, viz.: the gross sales less the expenses of running the farm, and (3) profits, which may be defined for the purpose of this discussion as the net proceeds less the interest on the investment.[A]

Assuming 160 acres of land, all tillable, devoted to dairy farming in eastern United States, gross sales may be estimated at $20 an acre, or an annual gross income of $3,200, and the net proceeds at $10 an acre, or $1,600. Under these conditions the young man's income would be $240, received as wages, plus $800, as his share of the net proceeds, or a total of $1,040 a year.

Generally speaking, probably a more satisfactory method, both for landlord and the farm manager, would be to pay the latter as nearly as may be what his services should be worth and give him in addition one-half the profits; that is, one-half of that which was left after deducting the expenses of running the farm and interest on the capital invested.

Merely for illustrating the method of calculation, let us assume this farm with its equipment to be worth $100 an acre, or $16,000. Let the farm manager be paid $840 a year. Assume the same gross income, $3,200, and the same cost of operating, $1,600, to which add $600, the additional salary of the manager. The total expense is then $2,200, and the net proceeds $1,000. If 4%, or $640, was charged on the investment, there would be $360 to be divided between landlord and manager, making the salary of manager $1,020. A simple calculation will show that if 5% were charged, the salary of the manager would be $940 a year, and if 6%, $860 a year. The advantage of the latter method of employment is that the young man runs less risk, while both receive equally any surplus beyond fair wages and fair interest on the investment.

In this connection it is important to consider how much may be reasonably paid for managerial ability. A study of the figures on page 133 will show that the labor income from a considerable number of farms of the better class was about 7% of the capital invested in the farms. The inference is, therefore, that if a man has $10,000 wisely invested in a farm he may pay $700 for a working manager; or, to put it in another form, before the owner of a farm can afford to pay $1,200 a year for a farm manager, he should have about $17,000 invested. Moreover, this investment must be in a form calculated to return an income. If part of it consists of investments for pleasure or fancy, such investment will not only not add to the income, but will detract from it by increasing the cost of maintenance.

This is scarcely less important to the employee than it is to the employer, since if the owner pays a higher salary than the manager can earn, he quite surely will sooner or later discharge his manager. This may result disastrously for the discharged young man, not merely on account of the loss of employment, but because his failure may militate against his securing satisfactory employment elsewhere. When an employer is seeking a man, he looks for one who has succeeded. There is an old saying, "Nothing succeeds like success," and it is only too true that nothing fails like failure.

[A] Profit is sometimes defined as that part of the product which the producer can consume without reducing his means of production.

CHAPTER III

FARM ORGANIZATION

In the last chapter were discussed the most common methods by which a young man acquires an opportunity to engage in farming. This chapter will discuss some less common arrangements by which may be bridged that period between the time the son is ready to go into the business and the time he may assume the complete control of the ancestral or other farm. It will also suggest a method for the continuous business management of a farm enterprise.

As stated, the most common reason for a farm changing from one family to another is the fact that no heir is willing to assume the obligation which is involved in paying for the interest of the other heirs. Connected with this problem is the further fact that the father is not usually ready to give up the management of the farm at the time one of his sons reaches the age to go into active business.

The reason for this state of affairs is made clear by the results of insurance statistics. The period that a man may be expected to live can be obtained by taking the difference between his present age and 90 and dividing the remainder by two. Thus, a young man who is 20 may reasonably expect to live 35 years, or until he is 55 years old. A man at 50, however, still has an expectation of life of 20 years, and the man of 70 of 10 years.

A farmer of 50 will usually have one or more sons ready to go to farming if they ever expect to engage in farming. But, as has been shown, a man of 50 has a reasonable expectation of 20 more years of life and cannot turn over the farm to his son, completely, without destroying his own opportunity for earning a livelihood. As things are usually arranged, therefore, there is no place on the average farm for the son, except as a hired hand, which is not desired permanently by either father or son.

Frequently the father fails to appreciate the earning power of his son, and, what is more important, that the boy has grown into a man. One day a teacher called a student of agriculture to his office, when the following conversation occurred:

"The Bureau of Soils at Washington," said the teacher, "has asked me to recommend several of our students to them for positions as field assistants. If you desire to have me do so, I would be glad to recommend you for one of these positions. The compensation is $1,000 a year and field expenses."

"I do not believe that I can accept," said Mr. Manning, "my father is in poor health and needs my help on the farm."

"Does your father want you to take charge of the farm and manage it so that you can make your training count?"

"No; my father expects to continue to manage the farm. He wishes me to work for him."

"How much does your father expect to pay you?"

"Thirty dollars a month."

The teacher found it extremely difficult not to interfere, but he merely said, "This is a case of filial duty which you must settle for yourself. I must have nothing further to say."

The young man returned to the ancestral home and is probably still there. It is, of course, impossible to determine the merits of an individual case, but this incident represents a type of cases where the son makes two important sacrifices from the sense of duty.

First, he sacrifices present, and, perhaps, future opportunity to earn the wages of which he is capable and to which he is justly entitled. And, second, and more important, he sacrifices the opportunity to develop his own powers and make concrete his own abstract self.

There are two things that every young man should do. One is to earn a living. A man that cannot or does not earn a living is of no value to himself or to anyone else. The other is to develop within himself his latent possibilities. He must apply himself to some problem, or problems, and through them develop his own personality. There is no place where more intricate and satisfying problems may be found than in the development of a successful farming enterprise. In the instance cited, the father may have been unable to pay his son the wage he might have obtained elsewhere, but he did not need to dwarf his son's development by treating him merely as a hired hand. His willingness to do so was probably due to his failure to appreciate that his son had become a man.

Sometimes a father is astute enough to reorganize his business so as to retain a place for himself while giving to his sons that opportunity which every man must have who develops himself normally.

An Ohio farmer once came to the Dean's office. He had a son in college who was just completing the first year of a two years' course in agriculture.

"I should like to have you find a place for my son in a cheese factory during the coming summer," said Mr. McKinley.

"I own a farm of 130 acres on which I have a herd of Jersey cattle," continued the father. "I have two sons and one daughter. I would like to have my sons about me, but there is no place for them on my farm because I am there and cannot get away. In fact, I do not desire to give up the management of the farm and the development of the herd of cattle."

"Not every father sees the situation as clearly as you do," interjected the Dean.

"This is my plan. After my son has spent a summer in a cheese factory, I want him to come back to your school for another year. I want him to learn, especially, all you teach about dairying. I will then build a cheese factory on my own farm and my son will make into cheese the milk of my own herd, and also from the herds of our neighbors. By the time he has completed his work with you, my younger son will have finished the high school. He has some liking for trading, and he will sell the cheese at wholesale and deliver it to the surrounding towns where markets are unexcelled. As for the daughter," continued this practical man, "she will get married and that will take care of her."

What became of the daughter is not known to the writer, but the rest of the program was carried out successfully and continued for many years.

A German came to this country and settled in New Jersey, where he established a large orchard. In course of time his two sons grew into manhood. While, of course, requiring plenty of laborers, the orchardist did not need the sons in the management of his farm. He, therefore, established one of these sons in the commission business in Philadelphia, thus, at least, keeping the profits on the sale of the products of his orchard in the family. He also needed cold storage for his fruit. The other son started a cold storage plant, which plays an important part in the profitable management of the orchard. Thus both sons have independent employment requiring managerial ability and the orchard is much more profitable than it otherwise would be.

Our land laws, our traditions and our practices are based upon the idea that a farm is to provide activity and support for but one family. In order, therefore, that the son may marry and begin to develop his life in his own way, it is essential to reorganize in some manner the method of managing the farm or to enlarge or, perhaps, specialize its activities. This may be accomplished on a simple partnership basis, or it may be in some such line as outlined in the illustrations which have been given. In other occupations such co-operative effort is the rule rather than the exception. That it is more difficult to effect satisfactory arrangements in farming must be conceded, else they would be more common. Doubtless it will often tax the ingenuity of father and son to devise the plans best suited to meet their particular problem.

There still remains to consider another form of business relation as applied to farming which has become almost universal in trade and transportation. The following incident may illustrate and emphasize the problem better than abstract discussion: One day a man walked into an office and stated that a friend had a half million dollars to invest in farming, provided that he could be convinced that the money would be invested profitably.

"Does your friend desire to buy land in any particular locality?"

"Yes," replied the promoter, "he wishes to buy land near ----. He has some sentiment about it. He was born in that neighborhood."

"Well, that is a rather bad beginning. Farming on sentiment is dangerous, especially when the sentiment is in no way related to the business."

The facts were that the region indicated was recognized to be one of the most unpromising sections of the state.

"If you undertake to invest a half million dollars in one neighborhood," continued the adviser, "you will pretty certainly fail to earn interest on your investment."

"Why?" inquired the promoter.

"Before you could possibly buy any considerable part of the land the owners of the farms you desire to buy would have doubled or perhaps trebled the price asked for their holdings. It is one thing to earn interest on an investment of $30 an acre and quite another to earn an equal per cent on $60 or $90 an acre.

"In the second place, farmers are content to accept less per cent on their capital than they would if it was loaned at interest, because the farm furnishes a home as well as a business. When you buy up all these farms and convert them into a single enterprise you will destroy their home value. You cannot hope to compete with the man, who, because his farm furnishes him a home, is content with an otherwise small return on his investment."

There were other reasons, of course, why such an enterprise would fail, which the speaker did not stop to explain.

"You are mistaken," challenged the promoter. "I intend to meet both your objections. My plan is to form a corporation and issue both preferred and common stock. The preferred stock shall bear 5% and that will belong to my friend who furnishes the money. I will retain the common stock. Five per cent is all the owner of the money is entitled to, while if the business returns more than that amount, it will be due to my management. I, and those associated with me, are entitled to all that is made above five per cent. By retaining the common stock the surplus income will come to us. Neither will I destroy the home value, because I shall associate the former owners with me in the conduct of the estate and may give them some of the common stock, so that they will be interested with me in making a profitable return. If they wish to keep their money invested in the farm, they will be given preferred stock in place of cash for their farms."

It is needless to say that the promoter never convinced his friend that he could successfully invest for him a half million dollars along the lines indicated. Nevertheless the corporate plan is not without merit. For example, if a father should incorporate his farm, he could provide for the inheritance of the preferred stock, among the heirs, as he desires. He could give to the son who operates the farm all the common stock, together with what preferred stock he is entitled or the father may desire him to have. The common stock would provide the means by which the income from the farm, which was due to the sons skill and management, might go to him. As time went on the son could acquire additional preferred stock from the father or other heirs, or he could invest his earnings elsewhere, as might seem most expedient. On the death of the parents, the preferred stock would be distributed as inheritance or the will provided without in any way interfering with the continuity of the farm enterprise. If at any time the son desired to discontinue the management of the farm, all he would need to do would be to dispose of his interest in the common stock at whatever he might be able to secure from the man who succeeded to its management. He could sell or retain his preferred stock.

Farming is the one remaining great industry that has not been organized so that a single enterprise may have a continuous existence. A corporation never dies, but at least three generations of men occupy the farms of the United States each century.

CHAPTER IV

OPPORTUNITIES IN AGRICULTURE

Some years ago, a prominent magazine contained an article entitled "The American Farmer's Balance Sheet," in which a descendant of the second and sixth Presidents of the United States was shown to have made in one year a profit of over $19,000 from a 6,000-acre wheat farm in North Dakota, and over $50,000 from a 6,000-acre corn farm in Iowa. A few months later there appeared in the same magazine another article, the purport of which was that great wealth, whether it be obtained from farming, the mining of coal, the manufacture of steel or the selling of merchandise, is the exception, while the man, in whatever calling, who rears and educates a family and at the same time lays by a small competence is the normal American product. The moral is that a $500-a-year-income farm is a more important factor to the national welfare than a $50,000-a-year-income farm.

In the latter article the writer tells of two brothers who had been reared on a Michigan farm. Reuben was tired of the country. He went to the city and apprenticed himself to a harnessmaker. Against the advice of young friends, Lucien bought sixty acres of land and ran in debt for it.

In a year Reuben was earning a dollar a day. He wore a white shirt and pointed shoes, not because they were more comfortable, but because other people did. He had no debts. Lucien had fair crops, but they yielded no more than enough to pay interest on the mortgage. He wore a ragged shirt, patched breeches and cowhide boots. People said that Reuben was making a gentleman of himself and learning a trade in the bargain.

In two years, Reuben had completed his apprenticeship. He was now earning $10 a week. He lived in a house that had a fancy veranda and green blinds. His clothing improved. Lucien was still ragged, but he paid his interest and $300 each year upon the principal. People said that Reuben, the harnessmaker, was bound to come to the front.

In ten years more, Reuben was still foreman of the shop at $50 a month. He lived in the same house, and smoked Havana cigars. Lucien built a new house and a barn. He smoked a pipe. The neighbors saw that every year he made some improvement on the farm. He wore a white shirt when he went to town, and he had a pair of button shoes. People said that Lucien was becoming a prominent man. His word was good at the bank.

Reuben began to complain that harnessmaking was too confining. His health was breaking down. The proprietor was selfish. He would not die and leave the business to him. Harnessmaking was not what it used to be. Lucien bought more land. He went fishing when he wanted to. Reuben came out now and then to spend Sunday. The birds seemed to sing more sweetly than ever before and the grass was greener. Lucien endorsed Reuben's note.

Lucien has pigs, and cows, and sheep, and chickens, and turkeys, and horses.

He raises potatoes and beans, and corn, and wheat, and garden stuff, and fruit. He buys his groceries and clothing and tobacco. Reuben buys everything. At the close of the year Lucien puts from $100 to $300 in the bank or takes a trip to Washington. Reuben does well if he come out even. Lucien does not fret; Reuben grumbles.

The picture is true to life. It has been enacted and re-enacted in every one of the older communities of the United States.

It has always seemed to the writer, however, that the author of this suggestive story left out two important personages. They were Sarah, the wife of Reuben, and Mary, the wife of Lucien. Sarah liked to make tatting and to go to pink teas. Mary preferred to raise flowers and fluffy little chickens. Nothing is to be said for or against the taste of either. Each has a right to her preference, but their point of view cannot be left out of the problem when a young man is considering his future occupation.

It has been said, and probably with considerable truth, that most congressmen would not hang around Washington if it were not for their wives.

No one must mistake this story as an attempt to compare harness making with farming, much less to compare living in the city with life in the open country.

What it does is to compare the struggle and the development of the man who goes into business for himself with the man who accepts employment at wages.

Because of less responsibility and less sacrifices at the beginning, the tendency is for young men to work for wages rather than to engage in business for themselves. This is becoming more and more true as industrial methods make it more and more difficult for the young man to command the requisite capital.

The man who works for wages usually has the larger income and appears the most prosperous during the earlier years as compared with his brother who enters business. The business man, however, who, while young, economizes and invests his savings in his business gradually outstrips his wage-earning brother. During later life he is able to enjoy the fruits of his earlier economy and investments, while failing powers and keen competition of younger and better trained men restrict the opportunities of the wage earner, who has generally spent his wages in better living, or at least in more outward show.

This is well shown by the fact that it is customary to make provision by means of pensions for wage earners of all sorts, while no such arrangement is made for men who engage in business, be that farming, trade or transportation.

For many reasons, however, young men will continue to seek employment at wages, even if only for a few years, or until some capital has been acquired which may be invested in business.

The question arises, therefore, what opportunities there may be for the young man who desires to engage, eventually, in the business of farming to work for wages along lines that will not be too far removed from the business in which he is subsequently to engage. It will be assumed that the young man has prepared himself in that same painstaking way that he would if he were preparing to become an engineer, a lawyer or a physician.

There is a constant demand for men with proper training as managers of farms. As stated elsewhere, the wages are seldom less than $40 nor more than $75 a month to beginners, although for men of experience $5,000 a year has been paid in exceptional cases for the management of large enterprises. These positions often constitute ideal opportunities for capable young men. They require, however, not only an intimate knowledge of farming, but the ability, also, to manage men.

The ability to manage men requires the combination of decision and tact, not possessed by all, and not easily acquired by education or practice. Not only must the farm manager be able to manage workmen, but oftentimes he must manage his employer, who may have little knowledge of farming but still insists upon having his own ideas executed, as he, of course, has a perfect right to do.

Another danger is the fact that where the farm is owned by a man engaged in other business, many circumstances may arise to cause the owner to change his plans or sell his property. There is often, therefore, a lack of permanency in these positions.

The United States Department of Agriculture employs upward of 5,000 people. There is a constant demand for young men to recruit this service, including experts in soils, plant production, animal husbandry, dairying, chemistry and forestry. Beginners receive from $800 to $1,000 a year. When they are sent out of Washington into field service, as many of them are, they receive their expenses, including subsistence in addition. Young men may rise rather rapidly by promotion to $1,600 a year, then more slowly to $2,000, while an occasional man is promoted to the more responsible position paying $3,000 to $4,000 a year.

The positions are all filled through the competitive civil service examinations. Examinations are held at more or less irregular intervals, usually several times a year, in various sections of the country. A letter addressed to the United States Civil Service Commission will secure the necessary information concerning openings and the general requirements for the examinations.

Employment in the United States Department of Agriculture often affords opportunity for varied experience and wide observation of farming methods throughout the country. Such employment is generally to be considered desirable if not continued for too long a period. As a matter of fact, men are constantly leaving the service to engage in practical or other work, a fact which makes the demand for young men greater than would otherwise be the case.

The various agricultural colleges and experiment stations are constantly seeking men. It would seem that the demand would eventually be satisfied. As a matter of fact, however, it grows greater year by year, both because these institutions continue to grow and because young men are attracted more and more to practical work. It is stated that in one institution there were 46 graduates in the course in animal husbandry and that 44 went into practical work and only two sought employment in college or station. The salaries are about the same as in government positions.

Agricultural newspaper work offers an attractive field for young men who are properly trained and have a taste for this kind of work.

There is also beginning to be quite a demand for teachers of agriculture in the high schools. As a rule a man is wanted who can teach, in addition, the sciences usually taught in secondary school. The customary salary is from $70 to $100 a month on an eight to ten months' basis. An experience of one or two years as a teacher in a high school, or even the lower grades of the public school, should be invaluable to the young man who expects subsequently to engage in farming. This is particularly true if he has not had the opportunity of a college training.

It is, perhaps, unnecessary to state that the salaries mentioned in this chapter are obtained only by young men who possess certain qualifications. To secure them, they must be men of ability, integrity, virtue and industry. No man who is not willing to make the preparation necessary to master his subject can expect to succeed. He must, also, be a man of absolute honesty, and he must lead a clean life. It was Bismarck who said, of German university students, "One-third die out; one-third rot out; the other third rule Germany." Every man who will may choose whether he will belong to Bismarck's second or third class.

The question for the young man of 20 is not merely as to the morrow, but what is likely to be the trend of events during the next 35 to 50 years.

"In 1800 the United States nowhere crossed the Mississippi and nowhere touched the Gulf of Mexico." In 1850 the country west of the Mississippi River was agriculturally largely an undiscovered region. Since 1870 we have much more than doubled our population and our agriculture. Since that time we have subdued more of the open country to the uses of man than we had been able to do in 250 years of our previous history.

During the past 300 years we have prided ourselves upon being an agricultural people. We have been an agricultural people, but our problems have not been chiefly those of the agriculturist, but those of the engineer.

Our problem, in the past, has not been to make two blades of grass to grow where but one grew before. Our problem has been to harvest and transport two bushels of wheat or two bales of cotton with the labor previously required to harvest one. Our crops have been so abundant that the agricultural problems connected with the growing of them has been secondary to the engineering problems of their harvesting and transportation. The self-binder and the steam locomotive have been our achievements.

If the writer mistakes not, the future problem will not be so much the harvesting and transporting, as the growth of the crops. In the future, young men will be needed who have studied the science of living things in order that they may make, literally, two blades of grass to grow where but one grew. To men who will be able to do so, will come success and honor.

CHAPTER V

WHERE TO LOCATE

Unless the young farmer expects to return to the ancestral home, the first question he must settle is where he is going to locate. Indeed, one of the most common questions asked is, What do you think of this state or that state or this region or that as a place to farm? There are few questions harder to answer. This is due, among other reasons, to the fact that every place has its advantages and disadvantages. The sum of the advantages may be greater in one place than in another, but if these advantages are known they must generally be paid for.

New adaptations, however, may change materially the value of the land in a given locality as, for example, the discovery that a region is especially adapted to raising alfalfa, onions, cabbages, apples or peaches. Changing conditions, as the growth of population or better transportation facilities, may materially affect the attractiveness of a region from the standpoint of the farmer.

The competition of other regions which grow similar crops is a potent factor in determining the desirability of a region. For example, the farmers east of the Allegheny mountains during the nineteenth century competed with the farmers of the central West who had free, fertile, easily tilled land on which to grow maize, wheat and oats. Cattle and sheep were pastured on the open range. The twentieth century has found the land of this region settled and capitalized in some instances beyond that of the eastern states; thus one factor at least of competition has been eliminated.

While farm values readjust themselves in time, it often happens, especially in the older settled regions, that farm values are slow in reflecting these changes in economic conditions. Changed conditions often call for a change in farm methods which the habits and traditions of even one generation prevent. To the man who is able to apply the proper methods the region may be a desirable one, although under existing conditions the results may be unsatisfactory. The young man, however, is cautioned at this point not to be overconfident of his own ability. Under such circumstances it is well to study the problem with great care, because the methods which seem unwise to the casual observer may, after all, be found to be based upon sound economic principles.

A man of 25 who is looking for a location should not only study the present conditions of the locality, but try to predict what is likely to be the future of the region during the next third of a century, since this is the period in which he may reasonably expect to be personally interested, although later in life he will find himself quite as much interested in the more distant future on account of his children.

Nothing is more self-evident than that one should choose a region, especially as regards soil and climate, which is adapted to the crop or crops to be raised, yet there are probably more failures due to a lack of crop adaptation than to any other cause that is not personal to the man himself. Not only do apples, for their best success, require certain soil types, but different varieties of apples require for their best development, distinctly different types of soil as, for example, Rhode Island Greening, Baldwin, York Imperial and Grime's Golden. Each reaches its best development on different types of soil and some require different climatic conditions. In like manner apples and peaches require distinctly different types of soil for the best success of each and for this reason peaches are not desirable as fillers in apple orchards.

If at the proper season of the year one goes from Pittsburg to Chicago via Columbus and Indianapolis, he will see great fields of winter wheat and a considerable number of permanent pastures. From Chicago to Omaha he will see only occasionally a field of wheat and scarcely any permanent pasture. Oats have taken the place of wheat. In parts of Eastern Kansas and Oklahoma the predominant crop is winter wheat. Throughout the whole region from Pittsburg to Topeka, Kansas, the characteristic crop is maize or Indian corn. Between St. Paul and Fargo, the main crops are spring wheat and oats. One may travel from Winnipeg, Manitoba, to Calgary, Alberta, a distance of over one thousand miles without seeing a field of maize. In some portions the main crop is wheat, in others it is oats.

These are illustrations of the crop adaptation over large areas, which has come about unconsciously, as has most crop adaptation. In other parts of the United States are to be found even more striking examples of crop adaptation, although the areas are much smaller, as in the case of tobacco, potatoes, celery, onions, apples, peaches and other fruits. Regions containing residual soils are more variable in crop adaptation than drift soils and require more careful watchfulness on the part of those who may wish to buy land.

As previously stated, advantages, if known, must usually be paid for. It comes about, therefore, that if a region or a farm is adapted to the raising of a certain crop which is more profitable than the average, such as maize, tobacco, alfalfa, celery, apples or peaches, this land will, other things being equal, command a higher price than land which does not possess this characteristic.

There is an underlying economic principle which the man who goes out to choose a farm should clearly understand. The principle has been stated by Fairchild as follows: "The normal value of products capable of indefinite multiplication tends always toward the value of least costly. On the other hand, if any production cannot be largely extended, so that the supply barely meets the requirements of the purchasers, the tendency of normal values is toward the cost of the most costly part of the product required to meet wants."

This principle explains why land especially adapted to raising maize is higher priced than land primarily adapted to raising wheat. Maize which enters into commerce is raised almost exclusively in ten states of the United States. Wheat is harvested practically every month of every year in different parts of the world. The young farmer should consider, therefore, whether he is undertaking to raise crops in which there is unlimited competition, or whether soil or other conditions cause the output to be relatively limited.

CHAPTER VI

SIZE OF FARM

The size of the farm is another of those questions on which there is endless debate and to which no general answer can be given. There are, however, certain rather definite principles which may help in settling an individual problem.

The size of the farm is related to the income per acre. If one's ideal or purpose is a gross income of $1,000 or $3,000 or $5,000 a year, he must consider how large a farm will be necessary to bring this return.

Assume, for the sake of discussion, it is desired to obtain a gross income of $4,000. In the eastern United States 200 acres of tillable land devoted to general farming may bring this amount. If the land is especially adapted to potatoes, and this crop takes a prominent place in the rotation, 100 acres might be sufficient to return the income named. Likewise a 100-acre retail milk dairy farm may produce a similar result. Forty acres devoted to truck farming or market gardening may be sufficient.

There is another way that the size of the farm needed may be estimated. There is a general relation between the gross income and the amount invested. In 1900 the gross income of the farms of the United States was 18 per cent of the total investment, which includes land, buildings, tools, and live stock. The average gross income varied for the different types of farming common to the northern United States from 16 to 19 per cent. This represents, of course, a great deal of very poor farming. The income of prosperous farmers must be somewhat better than this. If we assume that by careful methods the gross income is 25% of the total investment, then an investment of $16,000 will be required to bring a gross income of $4,000. While it is true that the gross income has no necessary relation to net income or profit, yet it is well to remember that a gross income is a necessary antecedent of a net income. The net profit from the production of a bushel of wheat, a dozen of eggs, or a pound of butter is of comparatively small consequence unless a sufficient quantity is produced.

A recent investigation by the Cornell station appears to show that with the type of farming now existing in Tompkins and Livingston counties, New York, where the investigation chanced to be made, the larger farms yielded the most profitable returns and that while present conditions exist, the size of farms is likely to increase rather than decrease. The fundamental reason seems to be the substitution of horse-drawn machinery for hand labor.

The following table shows the labor income on 586 farms operated by the owners, classified according to size:

Size (acres)	Number of farms	Average size (acres)	Labor income
30 or less	30	21	$168
31 to 60	108	49	254
61 to 100	214	83	373
101 to 150	143	124	436
151 to 200	57	177	635
over 200	34	261	946
----	----	Average 103	$415

While the larger the farm, the more prosperous was the operating owner or tenant, the size of the farm did not seem to affect the profit of the landlord.

The amount of land one individual may own is unlimited; the size of the farm unit is limited. After a farm unit has reached a certain size, depending upon the type of farming, the general arrangement of the farm and the skill in management, any further increase will increase the cost of operation, and as the increase continues eventually cause a decrease in profits. Assuming this to be true, it follows as a mathematical necessity that as the farm increases in size the total profits will increase as the farm increases up to a given point and then the profits will decrease. The following table illustrates this law:

Size of farm acres	A		B	
	Net profit per acre	Net profit per farm	Net Profit per acre	Net Profit per farm
160	$5.00	$800	$5.00	$800
200	4.50	900	4.75	950
240	4.00	960	4.50	1,080
280	3.50	980	4.25	1,190
320	3.00	960	4.00	1,280
360	2.50	900	3.75	1,350
400	2.00	800	3.50	1,400
440	1.50	660	3.25	1,430
480	1.00	480	3.00	1,440
520	.50	260	2.75	1,430
560	--	--	2.50	1,400

In both case A and case B it is assumed that the greatest net profit per acre is to be obtained with 160 acres, and that the net profit per acre when the farm is of that size is $5. In case A it is assumed that the net profit would decrease $1 for each 80 acres added, while in case B the decrease is assumed to be only one-half as rapid. In the first instance the net profit per farm increases until 280 acres are reached, when the net profit per farm decreases, until at 560 acres no profit would be obtained. In case B the net profit per farm increases until 480 acres are reached. Everyone is cautioned not to accept these figures as representing what would actually happen. All that can be said is that as the farm unit increases in size there will come a point at which the net profit per acre will decrease because of the physical difficulty of managing a large area, and, therefore, there is a limit to the size of a single farm. Fifteen thousand acres may lay in one tract and be owned by one individual, firm or corporation, but its economic management requires for purely physical reasons, not to mention others, that it be managed in several units more or less distinct from one another. Just what the size of this unit will be no one knows and it will vary with the type of farming, the type of farmer and many other circumstances. For example, a very common unit for a tenant cotton farm is between 20 and 50 acres, both the product and the farmer being a limiting factor.

Perhaps the most important lesson to be learned from a study of this table is that it is wise for some men to operate a farm of 320 acres, others of 160 acres and still others of 80 acres, because each size of farm presents a task suited to different abilities. It would be as futile for one fitted to operate only an 80-acre farm to attempt to manage 320 acres as it would be unwise for the man capable of conducting 320 acres to confine his attention to 80 acres. Unfortunately while this principle is not difficult to perceive and is easily stated, it is practically impossible to make any application of it to an individual case. Only time and the inexorable laws of competition will adjust men to their several tasks.

It will be of interest to note what influence in actual practice the type of farming has upon the size of the farm. The census reports the average size of all farms in the United States as 147 acres, with the different types as follows: Vegetables, 65 acres; fruits, 75 acres; dairy products, 120 acres; hay and grain, 159 acres; and live stock, 227 acres. Speaking in a very general way, only about one-half the land on these farms is in cultivated crops, while only 40% of the income may be from the products which cause the farm to be thus classified. The young farmer will do well to have these figures in mind when he starts out in life, for while they are not to be followed literally, they give him a measuring stick with which to compare his operations.

CHAPTER VII

SELECTION OF FARM

Having some of these preliminary questions settled, or at least well in mind, the young farmer is ready to inspect individual farms with a view to purchasing or renting. He should examine each farm from four general aspects, namely: (1) The character and topography of the soil, (2) the climatic conditions, including healthfulness and water supply, (3) the location, and (4) the improvements.

It may be well at the outset to emphasize the advantage which even a small difference in fertility may bring. Suppose one farm is capable of raising fifteen bushels of wheat per acre and another twenty bushels. If wheat is 80 cents a bushel, then the gross income is $12 and $16 respectively. If it is assumed that it costs in either case for seed, labor and interest on investment $8 an acre to raise and harvest the crop, then it will be seen that an increase of five bushels an acre doubles the profit. The comparison is perhaps not quite fair, since it costs slightly more to harvest the larger crop, but it serves to illustrate the point.

Neither the crop adaptation nor the crop-producing power of the soil can be determined by taking a sample and submitting it to a chemist for analysis. These factors can best be determined by the character of the vegetation, both domestic and wild, and by a knowledge obtained through observation or reading as to what this particular soil type usually does. Every type of soil has certain characteristics which under like conditions it may be expected to reproduce, much in the same manner as each species of animal reproduces its characteristics.

The first essential is to be able to recognize the different soil types. This can only be done by close observation and study. The second essential is to determine what the crop-producing characteristics of these types of soil are. This knowledge may be obtained by personal observation; but as most persons' opportunities are limited in this direction, it should be supplemented wherever possible by a study of the soil surveys of the United

States Department of Agriculture wherever these are available. When this is not possible samples of soil may be submitted to the Bureau of Soils of the United States Department of Agriculture or to the soil division of the state experiment station, together with a suitable description and such knowledge of the history of the land as is obtainable. In this way you may obtain information as to the natural adaptation of the particular type of soil.

There will still remain the question of the present condition of the land. For example, the Pennsylvania station obtained in a certain season 42 loads of hay from nine acres of land. The same season, from exactly the same soil type, the station obtained eight loads of hay from 20 acres. The condition of the soil was different, which the previous history of the two tracts of land fully explains.

It is of the utmost importance, therefore, to distinguish between the natural fertility of the soil and the condition of the soil. A further example will help to illustrate this point. At the Rothamsted Station a certain type of soil has for over 60 years produced annually about 12 bushels of wheat an acre without fertilizer, while with a complete fertilizer the same type has produced 30 or more bushels. The 12 bushels may be said to represent the natural fertility of the soil, while the additional 18 bushels may be said to represent the condition of the soil due to fertilizers or to other conditions. On the other hand, the natural condition of some other soil type might be only eight bushels, or still another type might be 16 bushels.

This principle is of considerable practical importance, especially in the eastern third of the United States. Generally speaking, clay and silt soils have a greater natural fertility than sandy soils; limestone soils than those that are deficient in lime. Thus soils that naturally grow chestnut trees, indicating a low lime content, have a tendency to deteriorate under exhaustive cropping much more rapidly than limestone soils. More fertilizers and other methods of soil improvement are necessary in the case of chestnut soils than in the case of limestone valley soils. One of the first questions to ask, therefore, concerning an unknown farm in Pennsylvania is whether or not chestnut trees grow naturally. It does not follow, however, that chestnut soils are undesirable. Much will depend upon the crop or crops it is desired to raise. For example, in some regions they are well adapted to potatoes and peaches. In these cases the cost of the fertilizers necessary to keep the soil in proper condition is small compared with the total return from the crop.

The pioneer's best guide as to the value of new land was and is the vegetation growing upon it, and, especially in a wooded country, the native trees. Basswood, crab apple, wild plum, black walnut, ash, hickory and hard maple generally indicate a fertile soil. White oak indicates only a moderate soil; bur oak, a somewhat warmer and better drained soil. Beech indicates a rather poor soil; a heavy clay, lacking in organic matter. Certain species of elms, maples and oaks, as red maple and the Spanish swamp oak, indicate wet soils.

The occurrence and vigor of certain herbaceous plants are especially indicative of fertility of the soil, as, for example, ragweed, bindweed, certain plants of the sunflower family, such as goldenrod, asters and wild sunflowers. Soils adapted to red clover and alfalfa are usually well drained and contain plenty of lime. Alsike clover will grow on a soil too wet or containing too little lime for either of the former. Soils that produce sorrel and redtop when red clover and timothy are sown need drainage or liming or both. Sedges usually indicate a wet soil, although certain species grow on dry, sandy soils. The point of this paragraph, however, is not to give comprehensive advice but to cause the young farmer to observe the conditions and make his own applications, which will vary in different regions and under different circumstances.

Perhaps the one feature that the young farmer is most likely to overlook in the selection of a farm is the relative proportion of tillable land. One farm of 200 acres, may, on account of stony land, wet land, comparatively unproductive woodland, or because of the arrangement of fences and roadways, contain only eighty acres of tillable land, while another may contain 160 acres. This is one reason why a 160-acre farm in the central West may be more valuable than a farm of the same size in the northeastern United States.

Columella says with regard to the selection of land that there are two things chiefly to be considered, the wholesomeness of the air and the fruitfulness of the place, "of which if either the one or the other should be wanting, and notwithstanding anyone should have a mind to dwell there, he must have lost his senses and ought to be conveyed to his kinfolk to take care of him."

In selecting a farm do not fail to inquire whether there has been any recent illness, and if so the nature of it, either among the persons living there or the domestic animals kept.

Aside from healthfulness, climate is a fundamental and controlling factor, both in productiveness and economic farm management. Temperature and rainfall affect the number of days that work can be performed upon the land and hence affect materially the economy of labor. It is this fact that prevents the systematic organization of labor so common in manufacturing and transportation. The climate also affects the cost of producing live stock by modifying the food and shelter required.

The climate of a region is best studied from the reports of the United States Weather Bureau rather than from the statements published by interested parties. So far as the production of crops is concerned the distribution of rainfall is more important than the annual amount, as may be shown by comparing the rainfall in such places as Columbus, Ohio, and Lincoln, Nebraska.

The average temperature during the growing season is, of course, of more importance from the standpoint of crop production than the average annual temperature. Maximum and minimum temperatures or the range of temperature must be considered as well as the average temperature.

One of the most practical questions to determine is the average date of the last killing frost in the spring and the date of the first killing frost in the autumn; in other words, the length of the growing season. Both altitude and topography enter into this problem. In a given locality killing frosts will occur on a still night in the valley before they do on the elevations, because the air as it cools becomes heavier and flows down into the lowest places just as water would do. On the other hand, as the altitude increases the growing season shortens.

Whenever I am asked a question involving the production of farm crops by a Pennsylvania farmer before answering, I ask three questions: (1) Where are you located? (2) Do chestnut trees grow naturally upon your land? (3) What is your altitude?

One factor that is often overlooked by the young farmer needs only to be mentioned to be thoroughly appreciated. It is the amount and character of the water supply. Not only is this of the utmost importance from the standpoint of the household, but it is fundamental to the best farm management. Thus, if the water supply is limited the amount of live stock kept will be curtailed, and thus the proper utilization of farm products prevented and maintenance of the fertility of the soil made more difficult.

The young farmer should recognize that some kinds of farming are more dependent upon the climatic conditions than others and should, therefore, select the location best suited to the type of farming desired or else modify his type of farming to suit the climatic conditions. If one studies critically the types of farming in various parts of the United States, it will be seen that they have already been adjusted in large degree, either consciously or unconsciously, to the climatic conditions. The young farmer should be careful that he does not undertake to butt his head against a stone wall.

Having found a farm that suits our ideal as to the natural conditions, such as the crop adaptation, fertility, topography and climate, what may be called the artificial conditions must be studied.

The location may be studied, both as to local and distant markets and the means of reaching each, which includes roadways and shipping facilities. Here again much will depend upon the products which are to be sold. The man who raises tobacco, hogs or beef cattle does not suffer any great economic disadvantage by living ten miles from a shipping station, but a man does who produces milk, peaches, potatoes or hay.

In these days there is not much danger that the character of the roadway will be overlooked by the intending purchaser of the farm, although sufficient importance may not be given to the advantage of really good roads, both as to grade and surface. Perhaps the one most important question to consider in connection with the transportation facilities is whether products may be shipped without change from the shipping station to the market it is desired to reach.

Although at first glance we may not like the thought, it must be conceded that neighbors are not only important morally and socially, but they also may have economic advantages and disadvantages. While it may sometimes happen that it will be wise to raise in a given neighborhood some product that no one else has undertaken to supply, yet as a rule, if a given neighborhood is raising Jersey, or Guernsey or Holstein cattle or Chester White, Berkshire or Poland China hogs, or Southdown or Shropshire or Cotswold sheep, it will be wise to raise the breed commonly raised instead of the least commonly raised breed, as it is sometimes supposed. The more potato growers or cabbage growers or celery raisers or orchardists in a locality the better for all concerned, for a number of reasons, among which may be mentioned (1) the more and the better the products raised the more buyers will seek the region and hence the higher will be the price obtained for the product; (2) the more of a given product there is to ship the better the shipping facilities for that product are likely to be; (3) all the necessary supplies for the type of farming can be more readily and cheaply obtained; (4) there will be a better knowledge of the business when more men have had experience in raising the particular crop.

These principles apply in all classes of business; thus we find woolen factories in Philadelphia, silk factories at Paterson, N. J., cotton factories at Lowell, Mass., plow factories at Moline, Ill., and steel mills at Pittsburg. Many of these centers possessed originally some natural advantages which caused the location of the first factory, but others have been drawn there on account of the principles enunciated. The farmers of a given region have a community of interest as well as railroads. The young farmer should recognize this fact and if necessary should exert himself to develop such interest in his community, both for his own benefit and that of his neighbors.

There are two classes of farms for which the purchaser is in danger of paying too much, one on which there are extensive improvements and one on which there are none at all. A farm with just barely enough improvements for the conduct of the type of farming it is proposed to develop can usually be purchased most advantageously. The purchaser should understand clearly that the previous cost of the improvements has no necessary relation to their present value, any more than the value of a second-hand suit of clothes is dependent upon its original cost. All depends on how badly they are worn and how well they are adapted to present conditions. The value of farm improvements is not unlike those in other business enterprises in this respect.

Their value depends upon present and prospective earning capacity and not on former cost.

No rule can be laid down as to the relation which should exist between the value of land itself and the value of the improvements. In practice it varies greatly. In the United States the farm improvements constitute on an average 21% of the total value of land, being as high as 45% in Massachusetts and as low as 15% in Texas. The young farmer may well consider, therefore, whether he can earn interest on his investment when the improvements cost more than 25% of the total value of the real estate. Certainly when it becomes one-half it is excessive. The man who runs a farm as an avocation usually errs in putting too much money into permanent improvements for the farm to be a paying investment.

If it is admitted that the farm unit is limited because of the physical difficulties of managing large areas, then it must at once be seen how important the arrangement of the farmsteading must be to the successful conduct of the farm. In the older farming communities where the present farm holdings are the result of several purchases or sales the shape of the farm, the arrangement of the fields and the place of the farm buildings become an extremely important matter. Sometimes satisfactory rearrangements are easily made, at other times they are quite impossible. No attempt will be made to discuss this subject in detail here, but the young farmer should bring to this question all the experience and study possible.

When the young farmer goes to inspect a farm it is to be assumed that he will be conducted over the farm by the owner or his authorized agent. It is proper to give respectful attention to everything that is told him, provided he follows carefully the California adage to "believe nothing you hear and only one-half what you see."

If a farm consists of 200 or 300 acres of land, it is possible for the agent to convey the purchaser over the farm in such a way as to prevent the least desirable portions being seen. If the farm has attracted the seeker of land, he should not purchase until he has made another visit, preferably some days or weeks after the first one. He may then very properly visit the farm alone, passing over quite a different course from that pursued hitherto. Sketches and notes will be found very helpful, and if the use of the soil auger is understood it may be well employed to study the character of both soil and subsoil. During the interval between visits some casual inquiries may be made among those who know the history of the farm in question, because the past history of the farm obtained from unprejudiced witnesses is of prime importance in arriving at a conclusion concerning its value.

A farm is much more attractive when a crop is growing upon it than when it is without active vegetation. Poor land looks relatively better than good land during or just after a rain. Many matters concerning the selection of a farm can only be learned by some years of practical experience. The young farmer will do well, therefore, to secure the help of some more experienced person. If he has among his acquaintances a successful farmer of mature years he will be fortunate if he can secure his advice.

CHAPTER VIII

THE FARM SCHEME

Farming is no pink tea. It is a serious business. After the young farmer has selected the farm he must develop his farm scheme. He must contemplate well and seriously the philosophy which underlies his plans. Unless he sees clearly what he is striving to attain and unless he understands the effect of his methods, he must fail in great measure to obtain his goal.

Satisfactory results in farming cannot be obtained as a general practice if the man is only interested in the results of a single year. For this reason the itinerant tenant system will not be satisfactory unless the landlord has worked out a satisfactory scheme which he requires his tenant to follow.

It is not enough that a man shall grow a single large crop, but it is necessary that he should continue to grow a satisfactory crop at least at regular intervals. For example, a piece of land may be adapted to cabbage, celery, potatoes or hay. Assume for the moment it is adapted to cabbage and that by one or more seasons of preparation an enormous crop of cabbages may be secured. This fact is of little value unless sufficient quantity is raised and the process can be repeated annually. Cabbages cannot be grown again on this particular piece of land for from four to six years on account of club root. If the farmer does not have other areas which he can bring into cabbages year after year, for from three to five years, then he becomes a failure as a cabbage raiser. Even a perennial, like alfalfa or asparagus, should form a part of the general scheme of crop production if the most satisfactory results are to be obtained.

There are two general questions at the basis of all farm schemes: (1) How to obtain a fairly uniform succession of cash products year after year, and (2) how to keep up or improve the fertility of the soil economically while doing so. In other words, how to keep the investment from decreasing while it is earning a satisfactory and fairly uniform income.

It is necessary, therefore, to consider what products are to be sold and what are simply subsidiary to the cash products. The cash products may, of course, be soil products or animal products, but more likely they will be both. When animals form a large part of the enterprise the cropping system must be carefully adjusted to meet the needs of these animals. Many apparently trivial details must be considered, as for example, whether the cropping system furnishes too little or too much bedding for the live stock.

In considering profits the enterprise as a whole must be kept in view. For example, if a man is producing milk, it may be cheaper, so far as the production of milk is concerned, to allow the liquid excrement to run to waste rather than to arrange for sufficient bedding. If, however, by using an abundance of bedding and saving all the high-priced nitrogen and the larger part of the potash in the manure, he is able to raise twelve tons of silage in place of eight tons, or three tons of hay in place of two tons, his enterprise as a whole will be more profitable when he uses the extra amount of bedding, although so far as the production of a quart of milk is concerned the cost is increased. It may be that by feeding corn to cattle or sheep one will obtain only 50 cents a bushel for his maize, while his neighbor is selling it to the elevator at 60 cents. If, however, the man who feeds his maize year after year thereby raises 60 bushels instead of 40 bushels, his enterprise, as a whole, may be more profitable than that of his neighbor.

As a matter of fact, the Pennsylvania experiment station has substantially these two conditions in certain of its fertilizer plats. When for 25 years the conditions have been similar to those where crops are sold from the farm, the yields have been: Maize, 42 bushels; oats, 32 bushels; wheat, 14 bushels;

and hay, 2,783 pounds per acre. But when conditions exist which represent the feeding of corn, oats and hay and the return of manure to the soil, the yields have been: Maize, 58 bushels; oats, 41 bushels; wheat, 23 bushels; and hay 4,190 pounds per acre. In the first instance the value of the products has been $15.75 an acre, while in the other case it has been $22.90 an acre.

Having worked out a cropping system that gives the proper yearly production of several crops desired, the next question to decide is how this cropping system and the disposition of the crops is going to affect the fertility of the soil. From a financial or economic point of view the most important soil element is nitrogen. First, because it costs from 18 to 20 cents a pound, while phosphoric acid can be purchased at five cents, potash at four cents; and, second, because of the readiness with which nitrogen may disappear from the soil under improper management, either through nitrification and leaching or by denitrification and passing back into the air.

Assuming a given type of management, the question is, How much of the required nitrogen will be obtained from the legumes in the cropping system, how much from the manure, and how much must be purchased in commercial fertilizers? No satisfactory cropping system can be devised at the present prices of farm products and cost of fertilizers for the production of the ordinary cereals and hay that does not include the production of some legume. Assuming a legume in the cropping scheme, the fertility of the soil may be maintained by yard manure alone or by commercial fertilizers alone. Illustrations of both methods are to be found in actual practice. Generally speaking, however, the use of yard manure supplemented with commercial fertilizers will be found more scientific and in the end the most economical.

A factor entering into this problem will be the amount of purchased feed. If considerable amounts of purchased feeds are used and the resulting manure carefully preserved and judiciously applied, the commercial fertilizers required will be reduced to the minimum.

A concrete illustration may bring out the philosophy underlying farm schemes better than abstract problems.

The following outline shows a five-course rotation with the method of fertilization which the results of the Pennsylvania Station indicated would be advisable, at least on limestone soils in eastern United States.

1. Maize yard manure, 8 tons per acre. 2. Oats nothing. 3. Wheat acid phosphate, 350 lbs. muriate of potash, 100 lbs. 4. Clover and timothy nothing. 5. Timothy nitrate of soda, 150 lbs. acid phosphate, 150 lbs. muriate of potash, 50 lbs.

This rotation is suggested for the purpose of maintaining a farm that is already in a fairly fertile condition and one on which there is no considerable amount of purchased feed. Where concentrates are purchased liberally, yard manure should be available to use on the timothy and meadow in place of the commercial fertilizers.

Where there is plenty of manure and it is desired to increase the amount of maize and hay and reduce the amount of oats and wheat, the following rotation and method of fertilization would be indicated:

1. Maize acid phosphate, 200 lbs. 2. Maize yard manure, 8 tons. 3. Oats nothing. 4. Wheat acid phosphate, 350 lbs. muriate of potash, 100 lbs. 5. Clover and timothy nothing. 6. Timothy nitrate of soda, 150 lbs. acid phosphate, 150 lbs. muriate of potash, 50 lbs. 7. Timothy yard manure, 8 tons.

Where there is plenty of yard manure, it would be also applied to maize under No. 1, or the yard manure could be applied to maize under No. 1, and commercial fertilizer applied to timothy under No. 6 could be repeated under No. 7. If the land is more or less depleted, an application of 200 pounds of acid phosphate to the oats would be advisable. However, the purpose is not to prescribe exact methods, but to point out underlying principles and their possible application.

As further illustration, it seems probable that the practice of a market gardener in using excessive amounts of stable manure might, in some instances at least, be modified to good advantage by reducing the amount of manure and increasing the amount of commercial fertilizer used. Unfortunately there is no experimental evidence bearing upon this question.

Potash required to maintain fertility is largely to be found in the coarse fodder, such as hay, maize stover and silage, and in the straw used for bedding; hence where these substances are used in abundance and returned to the soil the amount of potash required to be supplied in fertilizers is reduced to a minimum. Where, however, the amount of live stock is limited and the products sold contain large quantities of potash, such as hay and straw, the supply furnished in fertilizers must be liberal.

Phosphoric acid is always being slowly depleted from the soil either from the sale of farm crops or animal products. There is no way of returning this loss completely, except from the addition of a commercial fertilizer.

The above fertilizer suggestions are based on the experiments covering a period of more than 25 years on a limestone soil. Soils may modify materially the amount and application of the fertilizers, but not the principles enunciated. For example, a soil on which common red clover grows luxuriantly and has a prominent place in the farm scheme will require less nitrogen in commercial fertilizers in order to maintain the fertility than where legumes are raised with difficulty or do not form a part of the farm scheme.

One of the most important points to be emphasized is the fact that haphazard fertilization is not effective in maintaining soil fertility. If one starts out to establish a five-course rotation and build up his soil through a rational system of fertilization, he will obviously not obtain the full benefit of the rotation until he begins to get crops from the second round, which will be the sixth year from the beginning. It may happen, and unfortunately it has perhaps usually happened in the past, that during the first rotation the increase in crops has not paid for the cost of the fertilizers applied. In many instances a rational system of fertilization has not been introduced because the owner of the land could not afford to wait six years for his return. Profit in farming, therefore, does not consist in raising one big crop or even in obtaining a large balance on the right side of the ledger in a single year. It is both interesting and valuable to know that five tons of timothy hay, 45 bushels of wheat, 100 bushels of maize and 40 tons of cabbage may be raised on an acre, but the real profit in farming only comes through a lifetime of effort. To the man of capacity who prepares for his work the results will surely come, but they will not come all at once and, as in every other business, he must pay the price in hard work and close application to details.

In this connection it may be emphasized that one of the difficulties in successful farming is to find one man both interested and capable along the various lines essential to a successful farm enterprise. The danger is that a man will ride his hobby to the detriment of the other activities of the farm. A farmer friend of the writer, who keeps a horse and buggy, cares so little for a horse that for several years he has walked two miles each morning and each evening rather than to take the trouble to hitch up his horse. If one visits a high-grade breeder of dairy cattle, he is very apt to find his pigs of ordinary character. On the other hand, a specialist in hogs is likely to keep scrub cows. A man may be an excellent wheat raiser and a poor potato grower, and the reverse. The breeder of live stock is likely to be lacking in his methods of producing farm crops, while the up-to-date, so-called general farmer is not likely to be a special lover of live stock. In like manner, the man may be a successful farmer, dairyman or horticulturist from the producing side, but be a poor salesman. In fact, those qualities of mind and heart which make for the best success from the standpoint of production, whether soil products or animal products, is not that which makes the best trader.

It is not expected that the young farmer will be materially different from his hundreds of thousands of predecessors, but the better a man is trained and the more fully he studies his own adaptabilities and deficiencies, the more likely he is to succeed in the open country. For this reason, the young man should be careful to get as broad a training as possible. It is, therefore, often more important for him to study those things which he dislikes than to study the things for which he has a natural taste.

There was a man in our town And he was wondrous wise. He knew that if he wanted crops He'd have to fertilize.

"Its nitrogen that makes things green," Said this man of active brain; "And potash makes the good strong straw, And phosphate plumps the grain. But it's clearly wrong to waste plant food On a wet and soggy field; I'll surely have to put in drains If I'd increase the yield.

"And after I have drained the land I must plow it deep all over; And even then I'll not succeed Unless it will grow clover. Now, acid soils will not produce A clover sod that's prime; So if I have a sour soil, I'll have to put on lime.

"And after doing all these things, To make success more sure, I'll try my very best to keep From wasting the manure. So I'll drain, and lime, and cultivate, With all that that implies; And when I've done that thoroughly I'll manure and fertilize." Vivian

CHAPTER IX

THE ROTATION OF CROPS

The two essential reasons for a rotation of crops are: (1) The possibility of obtaining for the soil a supply of nitrogen from the air by introducing a legume at regular intervals, and (2) the prevention of injury to the crops from fungous diseases, insect enemies, weeds or other causes. Other reasons are often advanced, some of which are entirely erroneous, while others are of quite secondary importance.

The rotation should be carefully studied with reference to the farm scheme as previously outlined. Reasons for modifying the rotations are: (1) To change the kind or proportion of crops grown, (2) to change the amount of labor required, or (3) to increase the crop-producing power of the soil.

During 25 years the four crops of maize, oats, wheat, timothy and clover hay have been taken in rotation from the four tiers of plats at the Pennsylvania State College, so that the influence of the soil has been entirely eliminated. At the December farm prices for the decade ending December 1, 1906, the value of these four crops per acre have been: Maize, $29.67; oats, $14.49; wheat, $18.49; and hay, $18.05. It will be noted that during 25 years the average income from an acre of maize has been almost exactly twice that from an acre of oats. The region where these results were obtained is relatively unfavorable to a large yield of maize. It is obvious, therefore, that a modification in the rotation may modify the average income from the farm materially, provided such modification does not reduce the fertility of the soil. Thus, while the average income per acre during 25 years for the four-course rotation above mentioned was $20.17, if the rotation were increased to a five-course rotation by the addition of another year of maize, the average income would be $22.45 an acre.

It may be desirable to modify the rotation in order to increase or decrease a certain crop usually fed upon the farm. Thus, with a four-course rotation of maize, oats, wheat, clover and timothy, one-fourth the area would produce hay; while with a six-course rotation, composed of maize, oats, wheat, each one year, and hay three years, one-half the area would produce hay. If it is desired to still further reduce the area in oats and wheat, a seven-course rotation could be arranged with maize, two years in succession. This is the rotation that would be desirable for a dairy farm where it is planned to keep as many cows as practicable and to buy the concentrates largely. Either the wheat or the oats could be taken out of this rotation if either the one or the other were thought undesirable and a still greater amount of roughage desired.

On the other hand, there are places where the minimum amount of roughage is wanted. There are certain sections of the central West where it is possible to sow oats on corn stubble without plowing and where occasionally a rotation is practiced of maize, oats and mammoth clover. The clover is plowed for maize, the oats are disked in upon the corn stubble and the next year the clover is pastured until about June 1, when it is allowed to go to seed. In this rotation the only roughage obtained is the corn stover and the oat straw.

Another result reached by this rotation is that only one-third the land is plowed annually. In the four-course rotation mentioned above three-fourths of the land must be plowed, while in the six-course rotation one-half is plowed each year. In other ways the character of the rotation modifies the labor. For example, the labor and cost of harvesting an acre of hay is much less than that of producing, harvesting and threshing an acre of wheat.

Rotations may often be planned with reference to the main or cash crop. Thus in the Aroostook (Maine) potato district the rotation is potatoes, oats and clover. The chief purpose of the oats and clover is to keep down the blight in potatoes and add through the clover nitrogen and organic matter to the soil.

A system of cropping that is best when the owner operates the farm may not be desirable when the farmer is a tenant. When a farm is rented, the lease should provide that clover or other legumes occur with sufficient frequency to keep up the supply of nitrogen without the purchase of a considerable quantity in chemical fertilizers. The lease should be so drawn as to make it necessary for the tenant to keep live stock in order to realize the largest profit. The landlord should provide an equitable proportion of the mineral fertilizers when such are required.

The provisions of the lease and the character of the rotation will necessarily vary with circumstances, but the following system of tenant farming which has been employed for many years in Maryland will illustrate the principles just stated:

The lease provides for a five-course rotation consisting of maize, wheat, clover, wheat, clover. The landlord and the tenant share the maize and wheat equally, but the clover for hay or pasture goes entirely to the tenant, unless hay is sold, when it is divided equally. They each provide one-half the commercial fertilizer and one-half the seed, except clover seed, which the tenant is required to furnish.

This lease provides for two clover crops out of every five crops raised, thus supplying nitrogen abundantly, and the terms of the lease are such that it is necessary for the tenant to keep live stock to consume these clover crops in order to secure the most profitable returns. The feeding of the clover makes it necessary to feed some or all the maize and may lead to buying additional concentrates.

Stable manure is thereby supplied for the field which is to raise maize, while mineral fertilizers may be applied to the fields sown to wheat. On the limestone soils of the eastern states 50 pounds each of phosphoric acid and potash per acre applied to the wheat, and 10 loads of stable manure per acre to the maize will probably be found sufficient to maintain the crop producing power of the soil.

In laying out a farm for a rotation it is desirable to plan the number of fields or tracts that will go in a rotation and try to get these as nearly equal size as possible. Having decided upon the number of years the rotation is to run and having adjusted the fields or tracts accordingly, it is quite possible to modify the proportion of crops by adding one crop and dropping another at the same time. Thus, if there are six 20-acre fields, any one of the following rotations might be used and the change from one to another easily made:

1. Maize Maize Maize Maize Maize 2. Oats Maize Maize Maize Barley 3. Wheat Oats Oats Wheat Alfalfa 4. Clover and Wheat Clover and Clover and Alfalfa timothy timothy timothy 5. Timothy Clover and Timothy Timothy Alfalfa timothy 6. Timothy Timothy Timothy Timothy Alfalfa

During the first year the 20-acre field could be divided into four tracts of five acres each, containing potatoes, cabbage, tomatoes and sweet corn, and then followed for four or five years by any succession of crops above outlined. The point is that a definite adjustment of the farm to some general method of rotation and a definite system of fertilization and soil renovation do not prevent a considerable latitude in the crops raised. It will be obvious that the longer the rotation the more flexible it becomes in this particular, which is a point to be considered in laying out the farm and in adjusting fields and fences.

In some cases it may be desirable on account of the arrangement of the farm or the character of the crops to be raised to have two distinct rotations of crops. For example, if the farm lends itself to be divided into eight tracts, a five-course rotation of maize, oats, wheat, each one year, and clover and timothy two years, and a three-course rotation of potatoes, oats or wheat and clover may be arranged.

CHAPTER X

THE EQUIPMENT

The workman is known by his tools. The problem of obtaining the most efficient machinery for the conduct of the farm without having an excessive amount is not easy of solution.

It is probable that the cost of maintaining machinery and tools is not less than 15%, 10% for upkeep and 5% for interest, even under the most careful management. Doubtless in practice it is as much as 25%. If this is conceded there must be a limit to the amount which may be economically invested in equipment. This is a place where the lead pencil may be used profitably. For example, if $125 is invested in a self-binder, the annual cost of the machine at 15% will be $18.75. If one has but 15 acres of grain to harvest, it may be better to hire a self-binder at $1 an acre. On the other hand, it may be necessary to own a self-binder in order to get the grain harvested at the proper time.

Among the machines requiring a considerable investment for the number of days used may be mentioned hay loaders, hay tedders, corn-binding harvesters and lime spreaders. There is a certain class of labor-saving devices, however, for which there is more or less constant need, as, for example, means of pumping water, methods of handling manure, both from the stable to the manure shed, and from the manure shed to the field. This leads to the remark that there is at present great need of modifying our traditional ideas concerning farm barns. Why do persons usually sleep on the second floor, while horses and cattle are placed in the basement? Three things have brought about the need of a radical revision of our practices concerning the planning of barns: (1) Our present knowledge of the difference in the function of food in keeping the animal warm, and that of producing work, flesh or milk; (2) the discovery of the bacillus of tuberculosis; and (3) the invention of the hay carrier. It is not the purpose here to discuss barn buildings, but merely to call attention to the fact that the traditional barn has long since outlived its usefulness, and that the young farmer should plan his farm buildings to serve the purposes required in the light of modern knowledge.

Various attempts have been made to manufacture combined machines; that is, a machine which, by an interchange of parts or other modification, may be used for two or more purposes, as, for example, harvesting small grain and cutting grass. Such attempts have usually been unsuccessful. On the other hand, the young farmer should consider the range of usefulness of any given type of machine or tool; thus, a disk harrow is more efficient for some purposes than a spring-tooth harrow. For other purposes the spike-tooth harrow is better than the spring tooth. The spring-tooth harrow, however, will do fairly well wherever the disk harrow or the spike-tooth harrow is needed. When, therefore, only one of these tools can be afforded, the spring tooth may be a better tool to buy than either the disk or the spike-tooth, although it is not for certain purposes as efficient as either of the others.

The kind of machine should obviously be adjusted to the conditions, as, for example, the size of the farm, and the character of the farming. Riding plows may be desirable on level land, but where it is necessary to plow up and down hill, walking plows should be used. The extra weight of the wheel plow is not a serious matter on level land, because the sliding friction has been transferred to rolling friction, but no mechanical device has been or can be invented which will decrease the power necessary to raise a given weight a given height. The various machines requiring horse power should be adjusted, as far as possible, to require the same number of horses. If the main unit is three horses, then, as far as possible, all machines should require three horses, such as plows, harrows, manure spreaders, harvesters, etc. If the activities of the farm are sufficient to require six horses then some of the tools may require three horses each, while others require a pair.

[Illustration: Mr. R. H. Garrahan, Kingston, Pa., is one of the most successful growers of celery in the United States. After graduating from the Wyoming Seminary he spent one year studying horticulture at the Pennsylvania State College. For several years he was assistant in horticulture at the University of Tennessee. He now has at Kingston 60 acres under intensive cultivation. His principal crops are celery, asparagus, cabbage, tomatoes and onions.]

A farm with six work horses is rather a desirable one from several aspects. Among other things, it enables the farm owner to employ two men who can perform most of the team work with two three-horse teams, while at other times three pairs of horses may be arranged when the owner needs to use a team. This leaves the farmer time to attend to many activities not requiring horses, and time to plan the work and to look with more care after the purchases and sales. The size of such a farm will depend entirely on the nature of the activities. If it is a so-called general farm with a minimum of live stock, it would, perhaps, consist of from 150 to 180 acres of tillable land with some additional pasture and woodland. Ideally, every farm should have sufficient activity to make it something of a center. It should be an organism. It is difficult to organize one man.

It will be useful, when we come to discuss how profits may be estimated, to divide the capital into three general groups: (1) The plant, which in addition to the real estate, will include the machines and tools, horses used for labor, and other animals used for breeding purposes or for the production of animal products, such as butter, wool or eggs; (2) materials, which will include animals which are to be fattened for sale, and all seeds, fertilizers and foods intended to be turned into products to be sold; (3) supplies, which may include foods for teams, and money with which to pay labor, be this labor that of the farmer or his employees.

The purpose of this classification is to bring sharply into view the fact that the nature of different kinds of equipment varies. All the things named under the plant are in the nature of an annual charge against income. The charge under materials may or may not be an annual charge. If a man invests $2,000 in 50 head of cattle, which he intends to feed and sell for $3,250 at the end of one hundred days, he does not have to calculate interest on $2,000 for a year, but only for 100 days. Cattle paper is held in large quantities by banks in the cattle feeding districts of the United States. The farmer would, in fact, be unwise to keep $2,000 in the bank nine months in the year in order to use it three months. Like any other business man, if he has the money, he invests it and borrows the money to buy his cattle. The same thing applies to food and fertilizers. If the food is fed to cattle, some of the money invested in the food must pay interest during the fattening period. Food fed to dairy cattle and chickens may be paid for out of each day's income. In practice, the amount of money invested in food for dairy cattle and chickens is dependent only upon the most economical unit of purchase. One may apply fertilizers to buckwheat, give a three months' note for the fertilizer, and pay the note out of the proceeds of the crop. If the fertilizer is applied to one-year-old apple trees, this investment may be required to pay interest for fifteen years.

The same principle applies to supplies. If one starts into raising horses for sale, he needs to have some money or other income on which his laborers and his own family can live, say for five years, this being the age at which a horse is supposed to become salable. More people would raise apples and horses if they could afford to wait for the return on the investment.

While this is a serious handicap, it is an advantage to the man who arranges his farming methods so that he can secure an income from some other source in the interim. The young farmer will do wisely to so arrange his farm methods that a portion, perhaps the major portion of his farm, will give him quick returns while making some long-time investments, which later in life will give him a greater return because so few people are sufficiently forehanded to make them.

CHAPTER XI

HOW TO ESTIMATE PROFITS

No man who engages in manufacturing or merchandising knows how much he is going to make annually during life. Much less does he know how much he will be worth when he dies. Neither does the man who works for a salary or practices some profession for fees know what his annual income will be even during the following decade. Neither one nor the other knows whether he will die a millionaire or a pauper. It is a problem too complex for any human mind to analyze. It is less certain than what the weather will be on this day next year, because it is the resultant of more variable factors.

In some respects there is more hazard in farming than in manufacturing or in merchandising, while in other respects there is much less. The profit which may be obtained from farming is neither easier nor more difficult to estimate than is that of other commercial enterprises. However, there is no business in which more foolish estimates are made as to the probable profits, except, perhaps, in mining.

The purpose of this chapter is not to give advice as to possible or probable profits, but rather to point out the general character of the data required for any individual problem, where the data may be obtained and how it may be applied.

There are two forms or methods of stating the financial gain that has been obtained from farming or other business ventures during a year or other specific period. The first may be called the interest on the investment method, and the second the labor income method.

With the interest on the investment method, all expenses may be subtracted from all the sales. From the cash balance thus obtained the increase or decrease in inventory may be added or subtracted. This balance may then be divided by the capital invested, to determine the rate of interest received.

The rate of interest method is the usual method in the commercial world. The prosperity of the railroad or industrial concern is judged by the rate of interest it pays its stockholders on the par value of the stock. The stock itself takes on the capitalization in accordance with the present and prospective dividends. The fact that this method is generally used in the commercial world is evidence that it is well suited to its needs.

The young farmer who wishes to know whether the operation of a given tract of land in a certain manner offers him a worthy opportunity will not find the interest on the investment method the best suited for his purpose. This is especially true when applied to a single product. For example, it may be shown that 50 hens will, when properly managed, in connection with other farm enterprises, return a remarkable interest on the capital employed. It does not follow, however, that a man can make a living with fifty hens or even 500 hens. If a man has an investment of $5,000, on which he obtains 10 per cent, his income would be $500. If, on the other hand, he has an investment of $25,000 and obtains a return of only 6%, his income is $1,500, or three times the former amount. In neither case, however, does this form of statement tell a man how much of his income is due to his brain and brawn and how much to the capital invested.

What the young farmer wishes to know is how much will he receive for his own time, energy and skill, after deducting all expenses and a reasonable interest charge on his investment--such a rate of interest as he could get by placing his money in good securities or what he would be required to pay for his capital if he borrowed it. This is best obtained by the labor income method. With this method all expenses are subtracted from all sales and to the cash balance thus obtained is added or subtracted the increase or decrease in the inventory. This balance may be called the farm income. Thus far the procedure is just the same as the interest on the investment method. From the farm income is now subtracted a reasonable interest on the investment, the balance remaining is called the labor income. This is the return which the farmer has obtained by and for his own efforts. If this balance is zero, then he should change his methods or get into some other business.

This statement of his income, whatever it may be, enables him to compare his prosperity with that of the man who is employed upon a salary. Here, again, however, it is difficult to make comparisons because of the differences in expenses of living. The chief difference, however, in the expense of the wage earner in the city and the farmer is in the matter of house rent. For example, if the wage earner pays $300 a year house rent that must be deducted from his income in comparing it with the labor income of the farmer. It is often stated that the farmer also has his living from the farm. This was much more true formerly than it is at present. Under present methods of distributing food products and with modern types of farming, the amount of food supplied the table from the farm is comparatively small. The rancher in Montana eats foods canned in Maine or Delaware, while the New Hampshire farmer buys his vegetables from Boston commission merchants. The Minnesota farmer cannot supply his breakfast table with oranges, grapefruit or oatmeal. Many of them buy, if not their bread, at least their flour, and also their butter. The fact that the city man indulges in high living is no argument in favor of the country man expecting less wages. Some of those things which are necessary to make the country an ideal place to live are expensive. Some of them are more expensive to obtain in the country than in the city, as, for example, educational facilities. In justifying his purchase of an automobile, a young farmer recently stated that his wife had certain cares, responsibilities and even privations which her city friends did not have. He thought that the automobile would help to offset them.

To my mind there is no more ideal place to live and rear a family than in the open country when the conditions are what they should be and may be. I believe, however, it is well to insist that it costs something to live in the country as well as in the city if one lives as well as every farmer has a right to expect to live.

Let us now consider the steps necessary in order to arrive at a fair estimate of the labor income. To make the matter concrete, we will assume a farm of 200 acres worth $60 an acre located in central Pennsylvania on a limestone clay loam soil over 1,000 feet above sea level. This farm is to contain 20 acres of timber, a 30-acre apple orchard two years old, 40 acres of pasture, 96 acres of cultivated land divided into six 16-acre fields. The rest of the 200 acres consists of small yards, roadways and waste land. One-half of each of the six 16-acre fields is to consist of a rotation of maize, oats and wheat, each one year, and hay three years, the latter clover and timothy followed by timothy. The other half is to consist of maize, barley, followed by alfalfa four years. In the young orchard there will be grown for a few years potatoes, tomatoes, cabbages and garden peas. After the orchard attains a size which forbids these intertilled crops, a portion of the pasture may be broken up so that these market garden crops may be raised. There will be kept six horses, 20 milch cows, 20 ewes of some mutton breed of sheep, five brood sows and 50 hens.

First of all, let attention be called to the broad knowledge of farming required to operate this moderate-sized and comparatively simple farm. The crops to be raised are maize, oats, wheat, clover, alfalfa, timothy, potatoes, tomatoes, cabbages, garden peas and apples. The animal products sold will be chiefly butter fat, wool, mutton, veal, pork and eggs. This is neither a long nor complex list of products. They are all adapted to the farm which the writer has in mind. Yet the man who operates this farm to the highest success will need to have a knowledge of agronomy, or the raising of field crops, of horticulture, animal husbandry, including poultry husbandry and dairying. He needs to have a good understanding of the principles of agricultural chemistry, to have a knowledge of how to prevent and combat fungous diseases and insect enemies. To get the most out of his timber land he should know at least some of the first principles of forestry, and if he has gained some instruction in the study of landscape gardening, his home will be more attractive, and his farm a source of greater pleasure to him.

To proceed with the estimate, the first thing to be done is to make a record of the cropping system, giving the areas and the estimated production of each crop. How is the yield per acre to be determined? Clearly, one cannot afford to estimate his profits on the basis of some unusual yields. If one could be assured of 40 bushels of wheat, 60 bushels of oats, five tons of hay, 300 bushels of potatoes, or 200 bushels of apples per acre, or 500 pounds of butter fat per cow, or 150 eggs per hen per year, there would be no difficulty about obtaining a snug labor income. Such results are possible and are appropriate ideals for which to strive, but are not safe as estimates on which to do business.

The year books of the United States Department of Agriculture contain the annual estimate of the yields, and the average December farm price of staple crops by states. These figures may serve as a basis for making estimates. If the natural conditions are about the average stated, one may properly assume that he can obtain an increase of 50%. He may even hope to double the yield, although it is not safe to assume such an increase in making an estimate of profits. If the natural conditions are more favorable or less favorable than the average, he must take the fact into consideration in his estimates. In the same way he may consider whether the average December farm price represents fairly his expectation of the price, or whether because of favorable location or superior quality of the article purchased he can expect higher remuneration.

It is here assumed that the young farmer is himself going to be more than an average farmer. If he is not he will only get average results, in which case his labor income will be only that of the ordinary day laborer.

To repeat the idea in concrete terms. If the young farmer is located in central Pennsylvania and finds that the average yield of wheat for the state is 17 bushels an acre, he may safely estimate that his improved methods will bring him 25 bushels of wheat to the acre. He may even hope for 34 bushels per acre. At the Pennsylvania station several varieties of wheat have, during the past 18 years, averaged over 30 bushels per acre. One year one variety produced 43 bushels. It would not be safe, however, to use such figures in estimating profits.

Having outlined the cropping system and made a careful estimate of the total annual production of each crop, the next step is to determine the amount of food and bedding required for the live stock. From this data it may be determined what products will be available for sale, and what foodstuffs must be bought. Thus, it may be found, for example, that the amount of oats raised just meets the requirement, while more maize must be purchased, together with nitrogenous concentrates, and that a portion of the hay is available for sale. In the farm under consideration there will, of course, be wheat, potatoes, tomatoes, cabbages, garden peas and the animal products previously mentioned for sale, and later there will be apples and some lumber from the wood lot.

The data are now at hand by which to estimate the total receipts. Having made the estimates of receipts, the expenses are estimated, and the difference gives the cash balance, if there is any. The most important items of expense will be labor, feed, seeds, fertilizers, harvesting and threshing expenses, spraying material, shipping packages, blacksmithing and repairs. After all expenses that can be thought of are included not less than 10% should be added for incidental expenses.

The amount of commercial or natural fertilizers to be purchased is, of course, related to the yard manure which will be produced on the farm; therefore some estimate of the probable amount is desirable. In a roughly empirical way the amount of manure produced may be estimated at twice the amount of dry food and bedding used, provided it is hauled daily to the field. Where stored and drawn to the field at stated periods, the shrinkage in weight, although not necessarily in plant food, may be as much as one-half.

The estimate of what the inventory should be at the beginning and end of the year is not so simple a matter as it may at first seem to be. The purpose of taking the inventory is twofold: First, to determine whether the inventory has increased or decreased, and second, to determine on what amount of capital interest is to be calculated. For example, one must carry forward each year seed for the next year's crop. Feed must be carried over to feed live stock until other food becomes available, and there must be money on hand with which to pay for labor unless there is a cash income from the sale of products sufficient to care for the labor bills.

In the case of the farm under consideration there is a young orchard of about one thousand trees. This orchard is not bringing in any income, but there is a constant expenditure of money on it, and a constant increase in its value. While, therefore, it decreases the cash income it increases the farm income and the labor income. On the other hand, it increases the interest charges because the plant or farm is increasing in value. How much will it increase in value? In some sections it is customary to consider that an orchard increases in value $1 per tree per year. If this is a correct estimate, this 1,000-tree orchard will increase the value of the farm $1,000 a year until it comes into full bearing. The farm under consideration was purchased two years ago for $9,500. On the assumption just stated, at the end of 15 years from date of purchase this farm should be worth $25,000, at least $15,000 of which will be due to a 30-acre orchard. This is at the rate of $500 an acre for the orchard itself.

In order to bring out some of the phases of the inventory more clearly the following classification of items is given below:

INVENTORY

A. PLANT.

The real estate, 200 acres at $60 per acre. The live stock. Work horses and breeding stock. Machinery.

B. MATERIALS.

Seeds, potatoes, oats, maize, wheat. Feed, hay for cattle and sheep, silage for cows, maize for pigs. Growing wheat, 8 acres at $6 per acre. Live stock, calves, lambs and pigs.

C. SUPPLIES.

Hay and oats for horses. Money for current expenses.

In estimating the inventory at the end of the year, a deduction should be made for the decrease in the value of the live stock under the plant and also for the machinery. Perhaps 5% for the live stock and 10% for the machinery and tools will be a fair deduction. Under materials and supplies those items have been inventoried which are to be carried over each year from the preceding year. In the case of seeds the amount required must be deducted from the amount sold, or they must appear as a charge in the expense account. Ordinarily they are carried over from year to year and thus become a part of the permanent investment. Since on the farm under consideration there is a considerable monthly income from the sale of butter fat and eggs, it may be possible that no allowance will be needed in the inventory for current expenses, although it is always desirable to carry a bank account in order to be able to make favorable purchases when opportunity offers.

As a part of the work in a course in farm management, the writer asked each student to secure the financial history of an actual farm covering a period of three years. The financial history of 30 farms during the years 1901 to 1903, inclusive, and 28 farms during the years 1902-1904, inclusive, was thus obtained and is given herewith.

SUMMARY OF FINANCIAL HISTORY OF FARMS

Average size of farm, acres 143.21 133 Average area in crops (includes pasture), acres 121.1 112

Capital at end of three-year period $14,009 $8,893 Capital at beginning three-year period 12,962 7,704 ------- ------ Difference $ 1,047 $1,189

Interest on capital, $13,485, at 5 per cent[B] $ 674 $ 415 Increase in capital per annum 349 396 Average yearly receipts 3,613 2,208 Average yearly disbursements 1,907 1,221 Average yearly cash balance 1,706 987 Average yearly farm income 2,055 1,383 Average yearly labor income 1,381 968

These figures show the application of principles enunciated in this chapter. A careful reader will have no difficulty in recognizirg how the different items have been obtained. For example, the difference between the receipts and disbursements in the first column gives the cash balance of $1,706. The farm income, $2,055, is obtained by adding to the cash balance $349, which is the annual increase in the capital. The labor income is obtained by subtracting from the farm income the interest on the capital at five per cent. The amount of capital is determined by dividing by two the sum of the inventories at the beginning and end of the period.[C]

It will be noted that the gross receipts, the expenses, the farm income and the labor income on these actual farms are all more closely related to the capital invested than the size of the farm. Thus, on the 30 farms with a capitalization of about $13,500, the average yearly receipts were about $25 an acre, while on the 28 farms with a capitalization of about $8,300, the average yearly receipts were about $16 an acre. Lkewise on the high-priced farms the labor income was approximately $10 an acre, while on the lower priced ones it was about $7.

[B] Obtained by dividing by two the sum of capital at beginning and end of three-year period.

[C] For further details see Hunt, "How to Choose a Farm," Chaps. X and XI.

CHAPTER XII

GRAIN AND HAY FARMING

An important and primary factor in the production of all wealth is labor. Aside from the professional and domestic classes, the people of the world devote themselves to three forms of work: (1) Changes in substance, or natural products; (2) changes in form, or mechanical products; (3) changes in place, or exchange of products. The second of these forms of work gives rise to manufacturing; the third, to trade and commerce. Under the first subdivision two classes of natural products may be recognized; first, what, for want of a better name, may be called chemical products, such as ores, coal and salt, from which are derived mining and the metallurgical arts; and second, vital products, or, in other words, vegetation and animals. It is work applied to the production of vegetation and animals that gives rise to agriculture. Agriculture is labor applied to the production of living things.

KINDS OF AGRICULTURE

The industries which deal with the production of living things may be divided, theoretically, largely on the basis of the character of the results, but to some extent upon the nature of the activities involved.

{ Grain Farming--Cereals and } { grasses. } { } Agriculture { Plantations--Cotton, sugar, } { tobacco, coffee. } Plant Production { (Soil Culture) { Truck Farming, Market } { Gardening--Vegetables. } { } Horticulture { Fruit Growing--Fruits. } { } { Forestry--Trees, shrubs. }

{ Stock Raising--Work, meat, fats, hides. { Stock Feeding--Meat, fats. { Stock Breeding--Animals. Animal Production { Dairy Farming--Milk, butter and cheese. (An. Husbandry) { Sheep Husbandry--Wool raising. { Poultry Raising--Eggs. { Beekeeping--Honey.

Mixed Husbandry

The manner in which this theoretical classification has worked out in actual practice will be indicated in some measure by the inquiries of the United States Census Bureau. The twelfth census has classified farms on the basis of their principal income. If 40% or more of the gross income of the farm was from dairy products, it was called a dairy farm; if from live stock, a live stock farm; if from cotton, a cotton farm. If no product constituted 40% of the gross receipts, the farm was classified as a miscellaneous or general farm.

In 1900 there were 5,740,000 farms in the United States, which were, according to the rule just stated, classified as follows:

FARMS CLASSIFIED ACCORDING TO PRINCIPAL SOURCE OF INCOME

Kind of farm.	Total area, acres.	Number.	Average size per farm. acres.	Gross income acres.
Hay and grain	210,243,000	1,320,000	159	$760
Vegetables	10,157,000	156,000	65	665
Fruits	6,150,000	82,000	75	915
Live stock	335,009,000	1,565,000	227	788
Dairy produce	43,284,000	353,000	120	787
Tobacco	9,574,000	106,000	90	615
Cotton	89,587,000	1,072,000	84	430
Rice	1,088,000	6,000	190	1,335
Sugar	2,689,000	7,000	363	5,317
Flowers and plants	43,000	6,000	7	2,991
Nursery products	166,000	2,000	82	4,971
Miscellaneous	113,144,000	1,059,000	107	440
Total	844,000,000	5,740,000	147	$656

Including miscellaneous or general farms, there are just a dozen kinds of farms mentioned. Of this number, nine kinds obtained at least 40% of their products, and probably much more, from vegetable rather than from animal forms. However, live stock and dairy farms constitute about one-third of the total number of farms, and almost one-half the farm acreage. There are four kinds of farms on which the production of grain and hay forms an important part of their activities; namely, the hay and grain farm, the live stock farm, the dairy farm, and general farm. These constitute, in the aggregate, 75% of the farms of the United States, and by virtue of their larger area, they occupy 85% of the total farm area.

GRAIN AND HAY STATISTICS

At the close of the nineteenth century less than one-half the area of the United States was owned in farms. Only one-half of this farm area was considered to be under cultivation. The total area in cereals was one-tenth the total land area, while 3% was devoted to hay and 2% to all other crops except pasture.

Without going into details, it may be stated with reasonable assurance that: (1) During the last half of the last century, the production of cereals has increased much faster than the population. For example, in 1850, there were raised in the United States one ton of cereal grains per capita; by 1900 this amount had increased to one and one-half tons for each inhabitant.

(2) Since the number of persons engaged in agriculture has decreased in proportion to population, the quantity of cereals produced in proportion to persons engaged in agriculture has increased in still greater ratio. So far, therefore, as the amount of cereals is concerned, the farmer has been getting an increasingly larger return for his labor.

(3) The quantity of cereals has increased in proportion to the arable land. This may be due to one or more of three causes: (a) greater average yield per acre; (b) greater proportion of cereals to other crops; or (c) to a change in the ratio of the different cereal crops. The following table, giving the average yield of grain, reduced to pounds per acre, shows not only how the substitution of one cereal for another might affect the total production of cereal grains, but also suggests to the young farmer how he may modify the total product of his farm:

Yield Lb. Lb. in bu. per bu. per acre Maize 24.2 56 1355 Barley 23.7 48 1138 Rye 15.0 56 840 Oats 26.2 32 838 Wheat 13.2 60 792 Rice Paddy 746 Buckwheat 14.0 48 672

Yields will vary relatively in different regions and with different types of soil, and should be studied with reference to one's conditions.

(4) The wheat and oat crops have increased about six and one-half times in 50 years, the hay crop five and one-half times, while maize has increased four and one-half times. Cotton, the only other great staple crop, has increased four times in the same period. The oat crop has increased the most rapidly of any since 1880. It is interesting, and may be significant, to note that, while the production of wheat and barley in Great Britain has decreased about one-half in thirty years, the production of oats has increased somewhat.

(5) The greatest rate of increase in the production of cereals in the United States during the last half century has taken place since 1870. This increase is coincident with three other facts of the utmost importance: (a) The development of the central West, a treeless plain--prior to this period much of the farm land in the United States had been hewn out of the forest, tree by tree; (b) the consolidation of the steam railways into transcontinental lines; and (c) the introduction of the self-binding harvester. Formerly it took at least five men to do what is done today by one man in the harvesting of cereals.

ADVANTAGES OF GRAIN FARMING

(1) The cost of land excepted, the production of hay and grain requires a small outlay of money. During the past fifty years, many thousands of persons have been able to obtain farms of 160 acres at almost no cost. With a few hundred dollars invested in horses and tools with which to plow the prairie and sow the seed, these fortunate persons have oftentimes been able to pay the whole of their expenses, capital included, from the first crop. The renter who operates a hay and grain farm usually has but a small capital invested in his business.

(2) The cereals bring a quick return. Wheat may be sown in September and sold in July; maize may be planted in May and sold in November; oats may be planted in April and sold in August. The short period between seed time and harvest makes the oat crop a favorite one among renters. On the other hand, it takes from three to seven years to produce a marketable horse. It may take ten to fifteen years to begin to realize on an apple orchard.

(3) The products are not easily perishable, and hence can be held almost indefinitely. The development of the magnificent elevator system, based upon the principle that the cereals can be handled like water, greatly simplifies the holding and preservation of these staple products.

(4) The products are in constant demand, and hence they always find a market.

Agricultural commodities may be divided into three classes, depending upon the area which controls the price of the commodity, as follows: (a) price units world-wide, as wheat, cotton, pork; (b) price units local to large districts--

products too bulky to ship long distances--such as hay, potatoes and apples; (c) price units local to relatively small areas, such as strawberries and green vegetables. It is obvious that the larger the area which controls the price, the more constant will be the demand.

OBJECTIONS TO GRAIN FARMING

(1) It exhausts the soil. About two-thirds of the wheat of the United States is consumed outside the county in which it is raised.

(2) It requires a large quantity of land to produce a competence. Land must be low in price, or the interest on the money invested in the land will consume the profits. The relation of crop to income is suggested by comparing the gross returns from an acre of potatoes or tobacco with an acre of maize. The average gross income during a decade was, from an acre of maize, $9.50; an acre of potatoes, $38; and from an acre of tobacco, $61.50.

(3) Only such part of the land as is suited to tillage can be used.

(4) The marketing of cereals requires the transportation of bulky products. Hay is handicapped much more seriously. The distance a product can be shipped depends somewhat on the price per pound received for it. If it costs one cent a pound to ship maize to a grain market, obviously it cannot be transported without loss when it brings only 50 cents a bushel. On the other hand, two cents a pound may easily be paid for shipping butter which is worth 25 cents a pound. The transportation of $2,000 worth of maize to a railway station ten miles distant is a laborious and expensive operation, but when this same maize is turned into beef or pork, it will transport itself to the station with comparatively little trouble. Notwithstanding the excellent transportation facilities which the farmers of the United States enjoy, 80% of the maize is consumed in the county in which it is raised. Cereal production demands better transportation facilities than cotton farming, tobacco growing or the rearing of domestic animals.

(5) Capital must lie idle much of the time. The self-binding harvester or the hay rake is only used a few weeks, or perhaps more often only a few days, each year. A cream separator or a churn may be used every day in the year. In the first instance, there is not only interest on unemployed capital, but the

capital is actually deteriorating through nonuse.

(6) The production of hay and grain does not give continuous employment. The slightest consideration of the following table must show that unless live stock is kept, there are considerable periods of the year in which very little labor is required, while at other times considerable work is necessary to prevent loss.

TABLE SHOWING THE AVERAGE ACREAGE PER FARM OF PRINCIPAL CROPS.

	New York	Ohio	Wisconsin	Virginia
Maize	3	13	9	11
Wheat	2	12	3	6
Oats	5	4	14	1
Barley, rye or buckwheat	2	--	5	0
Hay and forage	23	11	14	4
Potatoes, beans or other vegetables	3	1	2	1
Fruits	2	2	0	1
Miscellaneous crops	2	1	0	2
Pasture, wood or unimproved land	58	45	70	93
Total size of farm	100	89	117	119

(7) Much depends upon natural forces. While there is opportunity for the use of knowledge and judgment in the production of high-grade seeds and even of large yields, there is not the same scope for skill that there is in some other lines of agricultural enterprise. Skill means the capacity to do something difficult, and the more effort required to produce an object the more value it has, provided its utility is unlimited. The farming which requires the most skill pays the best if one has the skill to apply to it. This is because those who do not have the requisite skill are usually unsuccessful.

CHAPTER XIII

THE COST OF FARMING OPERATIONS

Several millions of the inhabitants of the United States, not to mention those of other countries, are engaged each year in the preparation of the soil for the cereal and forage crops and on the work of seeding and harvesting them. The welfare of one-third the population is directly and that of the other two-thirds, although less directly, is quite as surely dependent upon the effectiveness of this effort. If, for example, as sometimes happens, one-third the population receives on account of untoward seasonal conditions but four-fifths of the usual product, everyone must suffer on account of this unrewarded labor. Many, perhaps most, financial panics have their origin in crop failures aided, doubtless, by an improper financial system.

Although widely and sometimes bitterly discussed, little is really known concerning the relation between the effort expended and the returns obtained in producing the great staple farm products; yet one of the most important and vital considerations in the organization of a farm enterprise is the income, both gross and net, which may be expected from the different crops contemplated. Obviously the yield and price of the several crops will vary with the locality and with the season. It is, therefore, impossible to predict for any year either what yield may be obtained or what price will be secured. If, however, a sufficient number of years are selected, an average may be found which will form a basis for calculating the probable result for another series of years. The following table gives the yield and the average farm values per acre for five staple crops for five years, 1905-1909 inclusive, for the United States and for four widely separated states, viz., Pennsylvania, Iowa, Texas and Oregon.

AVERAGE YIELD PER ACRE, 1905-1909.

Pennsylvania Iowa Texas Oregon Maize, bu. 36.6 33.4 21.1 27.3 Wheat, bu. 17.8 15.5 9.6 20.6 Oats, bu. 28.9 28.9 26.6 32.8 Potatoes, bu. 84.4 85.8 67.0 119.0 Hay, tons 1.39 1.56 1.32 2.11

AVERAGE FARM VALUE PER ACRE, 1905-1909

Pennsylvania Iowa Texas Oregon Maize $22.59 $13.80 $12.17 $19.58 Wheat 16.61 12.42 9.11 16.10 Oats 13.33 9.28 12.97 15.20 Potatoes 55.87 44.75 65.15 71.18 Hay 18.74 10.13 13.92 19.60

Such figures as the above may be compiled by anyone at any time for any year or series of years from the yearbooks of the United States Department of Agriculture. They form a fairly sound basis for calculating the gross income which may be expected from the staple farm crops, particularly for the cereals, potatoes, hay, cotton and tobacco. Five questions, however, present themselves, which should, as far as possible, be settled before applying them to an individual problem.

(1) How nearly do the conditions, especially those of soil and climate, of the

given location correspond to the averages of the state? The question can be settled only by a thorough study of soils and their crop adaptation. It is a matter requiring study, experience and judgment.

(2) How much larger yields may be expected on account of better methods employed? It is here that most mistakes are made in estimating possible farm profits. Necessarily, all statistical averages of production are much below those which an enterprising farmer considers an average crop and habitually produces. Not more than 50% increase upon these figures, however, should be anticipated by reason of the improved methods which one is going to employ.

While the average yield of maize, even in the so-called corn states, is not far from 30 bushels an acre, and while it is quite common for good farmers to produce 60 to 75 bushels of maize per acre, it would not be safe to assume a yield of more than 45 bushels unless the conditions are more than ordinarily favorable.

The application of the averages given on pages 149-150 to an individual farm enterprise may be illustrated by calculating the possible results which might be obtained on 80 acres of arable land in Iowa and Pennsylvania with the four great soil products of northern United States.

	Iowa		Pennsylvania	
	Acres	Income	Acres	Income
Maze	40	$552.00	15	$340.85
Oats	20	185.60	15	200.25
Wheat	5	62.10	15	249.25
Hay	15	151.95	35	655.90
Total	80	$951.65	80	$1,446.25

If 50% is added for the increased yields which may be expected on account of the employment of better methods, the total yield from 80 acres of arable land would become for Iowa $1,428 and for Pennsylvania $2,169. This does not mean that farming is necessarily more profitable in Pennsylvania than in Iowa. Not only may the cost of cultivating an acre of arable land be greater in Pennsylvania, but usually a larger territory must be owned in order to obtain 80 acres of arable land. Eighty acres of these four crops is probably as often grown on a farm of 100 acres in Iowa as on one of 160 acres in Pennsylvania. The total farm acreage in Iowa is, in round numbers, 35 millions; in Pennsylvania, 19 millions. In Iowa about one-half the farm area is in the farm crops under consideration, while in Pennsylvania these four crops occupy only one-third the farm area.

[Illustration: Mr. R. D. Maurice Wertz, after several years in railroad offices, took charge of his fathers farm at Quincy, Pa., in 1891, and converted it into a fruit farm. He now has about 220 acres in peaches and apples. It is understood that he has sent from the above shipping station and one other about $200,000 worth of fruit in the last six years.]

[Illustration: Mr. T. E. Martin, Rush, N. Y., is one of the most successful potato growers in the United States. He has a farm of 57 acres of the Dunkirk series of soil. He has three 18-acre fields in rotation consisting of potatoes, wheat and clover and alfalfa. Mr. Martin has increased the yield of potatoes from 60 bushels per acre in 1892 to 417 bushels in 1906. In 1906 he produced 7,510 bushels on 18 acres. In 1907 he sold $2,807.89 worth of potatoes from 18 acres, or $160 per acre. He attributes his large yields mainly to drainage, thorough preparation of the soil, good tillage, spraying, clover and alfalfa, manure and commercial fertilizers.]

(3) Will there be a general increase or decrease in the price of crops during the coming years?

The following table gives the average farm price for Missouri by five-year periods.

THE AVERAGE DECEMBER FARM PRICE BY PREVIOUS DECADES COMPARED WITH AVERAGE OF FIVE YEARS, 1906-10.

1866 1875 1886 1896 1906 to to to to to 1875 1885 1895 1905 1910 cts. cts. cts. cts. cts. Maize, bu. 40 33 33 35 49 Wheat, bu. 103 87 64 71 87 Oats, bu. 30 27 26 27 39 Potatoes, bu. 57 48 49 53 68 Hay, ton 902 799 704 700 875

An examination of the last column shows that the average price of these staple farm products has been considerably greater during five recent years than during the previous thirty years. Will this increase in price continue, or will there be a series of years of unusually low prices which will bring the average price of the decade down to that of the previous three decades? Few persons will care to venture an answer to this question, which is of the utmost importance to all farmers and especially to the beginner.

(4) The figures employed are taken from the yearbook of the United States Department of Agriculture and are the estimated farm price on December 1 of each year. Can the commodities be sold for the December farm price? Will potatoes sold at the time of digging bring less than the December price? Will wheat or maize held until May bring a higher price? To what extent, by the judicious holding of products, can advance in price be obtained?

(5) Will the products be sold for cash, or may they be turned into animal products at an increased profit? In some sections of the United States animals are reared primarily because of the increased profit due to manufacturing soil products into animal products; in other regions, however, they are kept primarily for the purpose of maintaining the fertility of the soil and only incidentally on account of the increased profits.

COST OF PRODUCTION

For a number of reasons it is difficult to determine the cost of growing farm crops. One reason deserves to be especially emphasized. In any business enterprise it may be necessary to run at a loss, because to stop would entail a still greater loss. This is particularly true in farming where men are employed by the month in order that they may be had when needed. Since they are receiving pay, it is better that such men should be employed some days at farm operations which return only a portion of their wages rather than not to have them employed at all. Under such circumstances, therefore, the cost of producing a given crop may be greater than is indicated by the time actually employed in its production.

Many other factors also enter, as the average number of hours per day which it is possible to work. This is greatly influenced by weather conditions. The Minnesota station determined that the working day on about thirty farms in that state varied from seven and one-half to eight and one-half hours, with two to three and one-half hours on Sunday. The average length of the working day for horses varied from 3.1 to 3.3 hours.

The cost for labor of cultivating a given area of land will depend not only on the crop or crops to be raised, the climate, the topography and character of the soil, the size and shape of the fields and the system of cropping, but also upon the man's ability for organization. It is said that the European farmers, and even the farmers from eastern Canada, are several years in adjusting themselves to farming in western Canada. When the farmers from Iowa, Kansas, Nebraska or surrounding states move into western Canada with their three-horse teams and other suitable equipment, applying their thorough knowledge of prairie farming, they are at once successful. The man is thus an important factor.

TIME REQUIRED FOR CULTURAL OPERATIONS

The following table will be helpful as showing time required to perform certain operations, since it is a record of labor actually employed on a field of 18 acres of easily tilled land in central Ohio. All labor was employed at prices named, board for man and food for horses being furnished in addition at the prices estimated. The owner of the land furnished the horse for the harvester.

Plowing 7.5 days at $2 $15.00 Harrowing 3 days at 2 6.00 Planting 2 days at 2 4.00 Cultivating (4 times) 7 days at 2 14.00 Cultivating with harvester 6 days at 1 6.00 Husking and cribbing by the job 45.54 Estimated cost of board 25-1/2 days 7.95 Estimated team maintenance 25-1/2 days 4.90 ------- $103.39

According to these figures the cost for labor of raising the crop and the cost of harvesting was almost exactly the same, each being a little less than $3 an acre.

THE COST OF PRODUCING FARM CROPS

The Minnesota station has determined the cost of growing the staple farm crops on 45 farms in different sections of the State. The total expense per acre for an average of six years is shown in the following table, not including land rental or cost of marketing.

COST OF PRODUCING FARM CROPS IN MINNESOTA.

Spring wheat, land fall plowed $5.54 Oats, land fall plowed 5.80 Barley, land spring plowed 6.89 Maize, husked from standing stalks 9.41 Hay, timothy and clover 3.68 Potatoes, land not fertilized 23.36 Potatoes, land fertilized 34.72

Some years ago the writer made an estimate of the cost of producing maize, oats, wheat and clover hay in a four-course rotation on a tenant farm in central Pennsylvania. The soil was a heavy clay and required plowing for each crop, except, of course, the hay crop, one acre a day being considered a good day's work.

Counting the expense of man and team at $2 per day, the labor cost per acre was found to be $7 for maize, $5.10 for both wheat and oats, and $2.30 for hay, or an average of about $4.90 per acre for the four crops. The interest on the capital invested in operating this farm, exclusive of the land, was estimated at $1.45 per acre.

INFLUENCE OF YIELD UPON THE COST OF PRODUCTION

The Illinois station has prepared a set of estimates upon the cost of producing an acre of maize, showing variations in cost due to differences in yield. In these estimates, instead of making a charge for the actual cost of manure or fertilizer applied, an estimate is made of the value of the plant food removed.

COST OF PRODUCING ONE ACRE OF MAIZE IN ILLINOIS AS MODIFIED BY YIELD.

Yield Yield Yield Yield 50 bu. 75 bu. 100 bu. 35 bu. Disking $0.40 $0.40 $0.40 $0.40 Plowing 1.00 1.00 1.00 1.00 Preparation .75 .75 .75 .75 Planting .15 .15 .15 .15 Seed .35 .35 .35 .35 Cultivation 1.00 1.00 1.00 1.00 Plant food 1.02 1.53 2.04 .71 Husking 1.25 1.87 2.50 .88 Marketing 1.00 1.50 2.00 .70 ----- ----- ------ ----- Cost per acre $6.92 $8.55 $10.19 $5.94 Cost per bushel .14 .11 .10 .17

The average yield per acre in Illinois for 12 years preceding date of this estimate was 35 bushels per acre; the average price per bushel during the same period was 32 cents.

LABOR COST OF PRODUCING A BUSHEL OF GRAIN

Not counting rent of land or interest on capital invested in equipment, nor depreciation of soil fertility, it has been shown that under favorable conditions, the labor cost of growing and harvesting an acre of wheat or oats may be as low as $4.50, and that of maize as low as $5 per acre. Assuming the average labor cost of producing an acre of wheat or oats at $5.50 and of maize at $6 per acre, and taking the average yields per acre for a series of years to be 13.8 for wheat, 30.9 bushels for oats and 24.9 bushels for maize, the average labor cost per bushel will be: Wheat, 40 cents; oats, 17-1/2 cents; and maize, 28 cents.

The data given in this chapter are to be accepted as suggestive rather than as determinative. The chief purpose in presenting them is to place before the young farmer an appreciation of some of the problems involved in the production of the chief and basic agricultural commodities. The young farmer's success will be modified by the role which they occupy in his farming system and by his ability to adjust them to the economic conditions in which he may find himself placed. A thorough understanding of the principle underlying the data submitted will go far toward enabling him to make this adjustment, although none of the illustrations given may have been obtained under conditions identical to his own.

CHAPTER XIV

THE PLACE OF INTENSIVE FARMING

The doctrine of the survival of the most fit applies equally to the field of biology and to the field of economics. The general introduction of vegetables and fruits into the human dietary has, by banishing the loathsome diseases of the Middle Ages, greatly increased human efficiency. It follows that those peoples or nations who employ vegetables and fruits in abundance, other things being equal, will be most fit to survive and must outstrip others less fortunately situated. We may for this reason alone look forward to the increasing importance of vegetable growing and fruit raising; but there is a more obvious and perhaps more direct reason. There is in the production of vegetables, at least, a method of satisfying the dietetic needs of an increasing population. The employment of a part of the area now in cereals and forage crops for the production of potatoes, cabbages, legumes, roots and tomatoes is one of the most ready means of increasing the food supply. Whether such substitution will be advantageous to the human race depends, however, not so much upon the food returns from a given area of land as upon the products from a given amount or unit of labor.

KINDS OF HORTICULTURE

In that form of intensive agriculture to which is given the designation horticulture, there may be recognized several more or less distinct divisions, as fruit growing, market gardening, truck farming and floriculture. Each has its own special problems, based upon conditions of culture and market. While, as in all classifications, there is more or less overlapping, the tendency is for them to become more and more distinct. The market gardener is the producer of vegetables for a local market, while the truck farmer produces similar products for a larger or wider distribution. The former grows a great variety of products, disposing of them in relatively small quantity, not infrequently directly to the consumer. The latter raises a few highly specialized crops which he sells in gross, usually through a commission merchant. Truck farming has developed since 1860, in consequence of the growth of large cities, which require enormous supplies of vegetables of fairly uniform quality, and on account of the continuous demand for fresh vegetables as nearly as possible throughout the year. Watermelons and sweet potatoes can be raised in the southern states and laid down in New York City or Boston more cheaply than they can be raised in the suburbs of these cities, and, what is equally important, they will be of superior quality.

The extension of railway facilities, the introduction of refrigerator cars and the building of cold storage plants has made it possible to grow in one climate products to be consumed in another. Cold storage has enabled the fruit growers of California to supply the eastern markets with peaches and other fresh fruit. Chicago, to give only one example, begins to receive strawberries, cabbages and tomatoes from the shores of the Gulf of Mexico early in the year and continues to receive these products, until finally they are being shipped late in the summer from the shores of Lake Superior. It is estimated that the change of locality from which these products come, travels northward at the rate of from 13 to 15 miles a day.

IMPORTANT FACTORS IN INTENSIVE FARMING

In the neighborhood of large cities, notably in the environs of Paris, market gardeners often produce their vegetables in made soil. The local character of the soil under such conditions is a matter of comparative indifference, since a board floor would answer every requirement as a resting place for the artificial soil. The large expense in preparing and constantly renewing the seed bed is only economically possible, however, where proximity to a large city out-weighs all other considerations.

Ordinarily climatic and soil adaptation are prime factors in successful horticulture--much more than in any other branch of agriculture. Each fruit has a restricted climatic range, and in most cases the number of soil types on which a given fruit can be made a commercial success is likewise limited. Thus, in general, apples and pears require heavier soils than peaches. Success in commercial apple growing requires even greater discrimination, since different varieties of apples demand different soil conditions. Thus Baldwins are grown the most successfully where a northern climate is modified by proximity to the Great Lakes. Rhode Island Greenings will succeed on soils too heavy for many other varieties. The York Imperial has not yet achieved a great commercial success save on one type of soil. Some varieties of apples are much more restricted in their adaptation than others. Thus, while the King is quite restricted, the Ben Davis has a fairly wide cultural adaptation. No one should plant an orchard until he has made a thorough study of his soil and climatic conditions and has received the highest possible expert assistance in choosing the varieties best adapted to his conditions.

There is an increasing tendency to specialize in vegetable growing. The production of celery, onions, muskmelons, watermelons, cabbages, cauliflowers, tomatoes and sweet corn, to mention only some of the most striking examples, are becoming more and more localized. Even where vegetables and flowers are grown under glass, not only is each house devoted to a single species, but, notably in the case of roses, growers are restricting themselves more and more to a few varieties. This is due to the fact that it is impossible to give in one house, or even in one establishment, the special set of conditions required for the most economic development of each species or variety of plant, just as in the open air the natural conditions are best adapted to a limited number of horticultural products.

So much being admitted, it follows that it is folly to attempt to grow plants under unfavorable climatic and soil conditions when competing in the same market with those possessing favorable ones. It is true, of course, that where one man fails another often succeeds, but this is no reason why a man should apply his talents under unfavorable circumstances. In fact, one of the important attributes of most successful men is their ability to recognize and apply their energies under conditions which will give them the most effective return for a given effort. There is no virtue in unnecessary toil. Progress in any enterprise, as progress in the human race, can be accomplished only in reducing the amount of labor required to produce a desired result.

All this is axiomatic. The purpose of emphasizing it here is that it is fundamental to the success of those who attempt to produce horticultural products. The necessity for the emphasis lies in the fact that these factors are so often disregarded. They are of most vital importance to the man who attempts to raise tree fruits. A mistake in the planting of celery, cabbage, or onions may be rectified the following season, but if a mistake is made in planting tree fruits, it may, as in the case of apples, require ten or even 20 years to discover the error.

The growth in commercial orcharding is due in part to the need of special knowledge and facilities for combating fungous diseases and insect enemies and to the better markets which a large production of uniform quality makes possible. While these are extremely important considerations, there is a more fundamental reason, which may in the long run exercise an even more potent influence. The location of the ordinary family orchard, so called, has been determined in almost every instance by the location of the farm buildings. There is no necessary relation between a good site for a farm dwelling and a suitable location for an orchard. It happens, therefore, that family orchards, taken as a whole, are not grown under as favorable conditions as are commercial orchards. This is a sufficient reason in itself, even if the other reasons above mentioned did not exist, why the commercial orchard must, in time, supplant these accidental plantings.

ADVANTAGES OF HORTICULTURE

The advantages of this intensive form of agriculture as compared with the more extensive forms discussed in Chapter XII may be stated as follows:

(1) A large gross income per acre may be obtained. An investigation of truck farming made some years ago indicated a gross return per acre about 40

times as great as that obtained on an average from all forms of agriculture.

(2) There is a large opportunity for the use of skill in raising and preparing products for market and an equal opportunity for the exercise of judgment in choosing the best markets.

DISADVANTAGES OF HORTICULTURE

(1) It requires considerable capital, particularly for machinery and labor. In the investigation in truck farming above mentioned the capital per acre invested in land, buildings, implements and teams was eight times that in the more general forms of agriculture.

(2) The products are for the most part readily perishable, requiring special facilities if held for any length of time.

(3) Growing out of above-mentioned fact, the market is easily overstocked at any given point, and hence prices often fluctuate widely.

(4) The yield is also quite variable, this class of products being especially influenced by seasonal conditions and particularly subject to insect attacks and fungous diseases. Since large capital is invested in labor, the horticulturist may be involved in financial ruin through causes which he is unable to control.

(5) The labor question, in certain forms of horticulture more than in others, involves difficulties, among which is need of large quantities of cheap labor for short periods of time.

CHAPTER XV

REASONS FOR ANIMAL HUSBANDRY

Animal products in the United States nearly equal in value those of all other farm products. Those soil supplies which constitute the food of domestic animals are not implied. Practically every farm in the United States keeps domestic animals, either for their labor or their products, and nearly every household in both city and country keeps one or more animals for companionship. The domestication of animals has been a prime factor in the civilization of the human race by furnishing man with motive force by which he has been able to increase his productive power; by giving him a larger, better and more regular food supply; and by furnishing the materials for clothing, making it possible for him to inhabit temperate and even arctic climates. Animals have not been less important in advancing the spiritual welfare of the human race, by inculcating habits of regularity and kindliness, which the care of domestic animals imposes.

INCREASE IN ANIMAL PRODUCTION

During the last half century animals have not increased in numbers as rapidly as have the inhabitants, but the value of animals has increased much more rapidly. While a part of this increase in value is due perhaps to a greater cost of production, a couple of illustrations will suffice to show that part of this increase in value has been due to increase in the individual merit of the animals. In 1850 sheep in this country produced 2.4 pounds of wool per fleece; in 1910 they produced 6.9 pounds per fleece. Thus, while in 50 years sheep have not quite doubled in numbers, the production of wool has increased more than five times. This is a striking example of the value of improvement in breeding, because the improvement in wool production is due to the influence of heredity in far greater degree than to the effect of improved feeding. Wool, like the hair on one's head, is not greatly influenced by the food supply, assuming it to be reasonably ample. Beef cattle offer another illustration of the way in which animal products have been increased without increasing the number of animals. Formerly beef cattle were matured in their fourth and fifth years, or even their sixth year. They are now placed upon the market in their second and third years. If animals can be matured in their third instead of their fifth year, t is obvious that a much smaller number of animals must be kept upon the farm in order to provide an equal annual supply for slaughter.

The increase in the size of our horses and the increased production of butter fat per cow which have occurred in the past half century are hardly less important factors in increasing the value of domestic animals and their products.

THE FUTURE OF DOMESTIC ANIMALS

One of the most striking features of recent progress in domestic animals is the large increase in the number of horses and the still greater increase in their value. There are those who have believed that the invention of many beneficent forms of mechanical power would in time, if not in the very near future, supplant the use of animals as a motive power. The fact seems to be, however, that they merely augment man's resources and increase his opportunities without lessening his need for animal power.

It appears reasonable to suppose that there will be witnessed in the United States a gradual shifting of live stock centers. During the past half century, the great central West has been noted for the production of live stock, particularly for beef, mutton and wool, as an incident of its pioneer development. Already the production of large herds of cattle and flocks of sheep has disappeared for the central West, and is now confined largely to Texas and the mountain states. The northeastern states are unrivaled in the production of grass, and have considerable areas less fitted for tillage than the prairie states. In time, therefore, the tendency will be for the regions best fitted to rear animals to increase their numbers of breeding animals. On the other hand, those states which produce grain in relatively large abundance may give more attention to fattening animals and to the production of dairy products which can be shipped long distances. As time advances, the history of other countries will doubtless be repeated. A greater distinction between the breeding and rearing of animals, and their fattening and preparation for market will occur.

ADVANTAGES OF KEEPING LIVE STOCK

Since animals occupy a place in practically all farm organizations, it is desirable to state briefly the advantages and disadvantages which may accrue to any individual enterprise. The most striking advantages affecting the farmer are:

(1) Animals make it possible to use land that would otherwise be wholly or partly unproductive. Hillsides and mountain slopes, soil too stony to cultivate, fields traversed by winding streams, and land partially covered with trees, are familiar examples. As previously mentioned, only about one-half the farm area in this country is improved land, and only two-thirds, even of the improved land, is in cultivated crops. The other third of the improved land

and a considerable portion of that half of the farm area known as unimproved land are utilized as pasture for domestic animals.

(2) They make use of farm crops which would be entirely or partially wasted. Straw, the stalks of maize, clover and alfalfa hay and other leguminous forage crops would not have sufficient value to pay for raising if animals were not kept to convert them into useful products. In fact, the usefulness of a given animal may be judged by the economy with which he converts these otherwise useless products into food or other materials for the use of man. The most profound studies are being made to determine the conditions under which this takes place.

(3) In thus acting as machines in manufacturing raw materials into finished products animals convert these coarse and bulky materials into those which are much more concentrated, thus making their transportation economically possible. A pound of beef has required food containing ten pounds of dry substance, and a pound of butter has required thirty pounds of dry matter to produce it.

These refined products may be shipped around the world, while the raw materials may not be profitably transported beyond the county in which they are raised. Moreover, the farmer has the profit which comes from manufacturing the raw materials into refined products.

(4) In the production of these finer products much of the essential materials of plant growth are left upon the farm. The experiments of Lawes and Gilbert show conclusively that in fattening animals more than nine pounds out of ten of the essential fertilizing ingredients of the food reappear in the solid and liquid excrements. Prothero says: "Farming in a circle, unlike logic, is a productive process."

The fiscal policy of one of the great nations of the globe is based upon this idea. Everything possible is done by Germany to encourage the keeping of live stock, because the more live stock that is kept, the more productive will be the soil. The larger the crops raised the more people will be required to harvest them and the larger will be the population to recruit the army and navy. The Kaiser and the German scientist recognize that the fighting force of the Empire is related to the number of domestic animals reared. The meat supplies of the people are, therefore, taxed to bring about this end.

(5) The rearing of live stock makes it possible to arrange a better rotation of crops. A five-year and, even better, a six-year rotation, is more effective than a four-year in maintaining the crop-producing power of the soil and enables the farmer to reduce his cost of production. It is possible to keep a larger proportion of the farm in grass and other forage crops, thus reducing the amount of land plowed annually and at the same time decreasing the exhaustion of the land, provided the forage crops are fed to live stock upon the farm.

There is an old Flemish proverb which reads:

"No grass, no cattle; No cattle, no manure; No manure, no crops."

The point of this proverb is that good grass is the basis of good agriculture. Investigations have shown that one may go farther and say that one of the most ready means of increasing the crop-producing power of the soil is by adding fertilizers to grass land. The large number of plants per acre enables the plants to utilize the fertilizer to the highest degree, and plowing under the resulting dense sod is one of the most effective methods of enriching the soil.

(6) Animals require constant care, thus making possible a more constant use of labor and other capital. The wheat farmer of North Dakota sows his wheat in April and May and harvests it in July and August. He usually threshes it immediately, and is practically without employment for himself, his teams or his men from September until April. On live stock farms the labor employed in the summer in the field is needed in the winter in paddocks and stables.

(7) The management of live stock, including the rearing of poultry and the manipulation of dairy products, may be made to require a higher skill than the production of farm crops as ordinarily practiced. The communities which have given the most attention to dairying and to the rearing and fattening of animals have generally been the most prosperous.

DISADVANTAGES OF KEEPING LIVE STOCK

(1) Keeping live stock increases the capital required to operate a given area

of land, especially where animals are kept in connection with the production of hay and grain. Not only must there be capita with which to purchase animals, but usually more is invested in buildings. In a self-contained farm-- that is, one which raises sufficient food for the requirements of the live stock- -ten dollars an acre may be considered a moderate investment for animals. If, however, the plan is to raise only the coarse feed, while the necessary grain as well as other concentrates is largely purchasec, a farm may easily carry from $25 to $35 worth of live stock per acre. Lack of capital is one of the most potent influences in preventing a larger production of animals and animal products. Cattle paper, or notes given to secure money for the purchase of fattening animals, is a common bank asset in the feeding districts of the central West.

(2) The very perishable nature of animals entails a great risk in the investment of capital in live stock. Not only the products of a single year, but the growth of a number of years, may be suddenly swept away by disease. This may include the crops of several years, thus destroying capital invested in the production of the crops as well as the capital originally invested in the animals. Many a farmer has seen the gradual accumulations of years rapidly melt away in the presence of some contagious disease. Tuberculosis in cattle, cholera in hogs and liver rot in sheep are striking examples of diseases that have caused the farmers of this country untold losses.

(3) When an animal has been properly fattened he must be sold. If held for any great length of time, not only is there a constant outlay for food to maintain the animal, but the condition of the animal may actually deteriorate. Hence it is not possible to hold animals for a better market for a long period of time, as is possible in the case of the cereal grains.

(4) Serious losses may occur where profit was expected through a rise in the price of foodstuffs. Scarcity in food supplies, due to an unfavorable season, often compels the stockman to sacrifice animals that he has been raising for two or three years. It is sometimes asserted that, although society suffers from short crops, the farmer is benefited, because the increase in price is greater than the decrease in yield. One year, for example, the decrease in the production of maize was 30%, while the increase in price was 50%. If, therefore, the crop had been sold it would have brought more than the crop of the previous year. The farmers, however, require about 80% of the maize

crop in the production of their live stock, so that when there was a decrease of 30% in the yield of maize, many had none to sell, while others had to purchase maize at increased prices or use other crops, such as oats, which they might otherwise have sold. Still others would be compelled to sell, at reduced prices, their partially fattened animals. There is a constant fluctuation in the price of animals and animal products, due to variation in yield and hence in price of food supplies. It requires continual vigilance on the part of the stockman to secure food supplies at such cost as will enable him to secure a profitable return from his animals.

CHAPTER XVI

RETURNS FROM ANIMALS

In any well-considered plan of farm operations it is essential to have some basis for estimating the amount of food required to carry live stock through the year in order to know, on the one hand, what portion of the crops raised are available for sale and, on the other hand, what food supplies must be purchased. A requisite of any successful farm enterprise is a proper consideration of these market conditions. While domestic animals consume a variety of foods, and each class of animals has special food requirements, the basis of calculation of the needed supplies is fortunately not complicated. Twenty-five pounds of dry matter are required per day for each thousand pounds of live weight of horses, cattle and sheep, and for swine about 40 pounds for each thousand pounds of live weight. It may be more convenient to calculate the food requirement of swine on the basis of increase in live weight, allowing five pounds of dry matter for each pound of increase. Some further details as to food requirements will be found in the paragraphs which follow.

COST OF PRODUCING HOGS

Pigs possess two characteristics which make them unique among domestic animals. They consume concentrated and easily digested foods only, and they produce nothing but meat, fat and bristles. Cattle furnish milk and hides; sheep, wool, hides and sometimes milk; fowls furnish eggs and feathers. On account of their limited range of usefulness and because of the high value of much of the food consumed, it would not be possible to rear swine economically were it not for their prolificacy and the fact that they are employed largely as scavengers. Many cattle are fattened without direct profit. The indirect profit comes from the sale of the pigs which have followed the cattle. It is customary to mature one hog with little or no additional food while fattening two steers. In many well-known ways, pigs consume products which would otherwise be wasted. This is especially true in the more densely settled sections of the world.

On account of their prolificacy, the returns obtained for the amount of capital invested is greater than in the case of sheep, cattle or horses. Ten sows, worth $100 to $150, are sufficient to produce 100 pigs; 75 to 80 ewes, worth from $300 to $500, are required to produce an equal number of lambs; 110 cows, worth $4,500 to $6,000, to produce 100 calves; and 200 mares, worth from $20,000 to $30,000, to guarantee 100 foals. To put the matter in another way, the capital invested in swine may be reproduced in the offspring ten times in one year; the capital invested in horses not more than once in five years.

In general, 500 pounds of maize will produce 100 pounds of pork, which is equivalent to eleven pounds of pork from a bushel. Since hogs are so largely produced from maize, the price of maize and the price of pork are very closely related. For example, if maize is worth fifty cents a bushel, the grain required to produce a pound of increase in live weight will cost about 5 cents; if 40 cents a bushel, 4 cents; if 30 cents a bushel, 3 cents; and so on.

COST OF PRODUCING SHEEP

In the classic investigations by Lawes and Gilbert, food containing 100 pounds of dry matter produced a live-weight increase of nine pounds in steers and 11 pounds in sheep. At the Wisconsin station, sheep required less food than steers per pound of gain. During rapid fattening of sheep 500 pounds of clover hay and 400 pounds of maize may produce 100 pounds of increase in live weight. While swine require a less weight of food for a pound of increase than sheep, on account of the more digestible character of the food eaten, yet the Wisconsin station found that the expense of producing a pound of increase was less in sheep on account of the less expensive character of the food.

MEAT AND MILK PRODUCTION COMPARED

A summary of the investigations of American experiment stations shows that 100 pounds of dry matter produced ten pounds of increase in live weight of steers. The same quantity of food when fed to milch cows produced 74 pounds of milk, plus one pound of increase in live weight. This 74 pounds of milk contained 3-1/4 pounds of fat. In general, therefore, the food required to produce a pound of butter fat is about three times that required to produce a pound of increase in steers.

COST OF STEER FEEDING

The fattening of beef animals is largely conducted by farmers who make a specialty of it. This is particularly true in the so-called corn belt. Into this region are gathered the two and three-year-old and, more rarely, yearling steers, many of which have been reared in Texas or in the mountain states where the supply of maize is not sufficiently ample to fatten them. These are placed in paddocks with open sheds, where they are fed from 90 to 150 days, after which they are sent to market for slaughter. The food consists usually of maize fodder, maize stover, hay, maize (usually in the ear), a little bran, linseed or cottonseed oil meal. The ration per day during rapid fattening is about 20 pounds of dry matter per 1,000 pounds of live weight, containing 16 pounds of digestible substance, of which 1.25 to 1.75 is digestible protein. One hundred pounds of increase may be obtained under average conditions from 150 pounds stover, 325 pounds of hay, 775 pounds of maize and 75 pounds of cottonseed meal.

Great variations will occur, however, depending upon the condition of the animals at the beginning of the feeding period and the degree of fatness or finish to which the animals are brought before placing upon the market. In any case, the food consumed will cost more than the value of the increase. The only way that steers can be profitably fattened is by increasing the value per pound of the animal. Thus an 800-pound steer may be purchased at five cents per pound, or $40. After feeding, say 150 days, he may weigh 1,100 pounds, when to bring a profitable return he should sell for 6 cents a pound, or $65. This is a gain of $25, eight of which came from the increase in value of the original 800 pounds. Usually steers cannot be fattened profitably unless there is an increase of at least three-quarters of a cent per pound in the value of the animals and then, as previously explained, only in connection with the hogs which follow them.

COST OF PRODUCING MILK AND BUTTER FAT

Well-selected and properly fed cows may produce 240 pounds of butter fat annually. The amount of fat obtained will depend upon the richness of the milk. Thus, 8,000 pounds of 3% milk, 6,000 pounds of 4% milk, or a trifle less than 5,000 pounds of 5% milk, will give this quantity of butter fat. These are customary returns from different types of cows.

If each cow in the herd is dry for six weeks each year the daily average of the cows actually milked will be three-quarters of a pound of butter fat. There are herds which make an average of nine-tenths of a pound of butter fat per day, but to secure this result requires superior cattle, careful feeding and more than ordinary care.

The standard ration for milch cows weighing from 1,000 to 1,200 pounds is 25 pounds of dry matter, two-thirds of which is digestible. The ration should contain not less than two pounds of digestible protein. In ordinary practice, about ten pounds of the dry matter of the ration is obtained from maize silage, nine pounds from hay and about six pounds from grain or other concentrates. In general, this is obtained by feeding 35 pounds of maize silage, ten pounds of hay and seven to eight pounds of concentrates. The silage may be estimated at one-tenth to one-eighth of a cent a pound, hay at from one-fourth to one-half cent and concentrates at from three-quarters to one and one-quarter cents per pound, varying, of course, with the different sections of the country. The amount of food needed will vary somewhat with the size of the animals, but will depend much more largely upon the amount of milk and butter fat given. While maintaining substantially the general average just given for the whole herd, it is the practice of careful feeders to vary the amount of concentrates fed to each individual in accordance with the amount of butter fat or milk given.

[Illustration: Mr. Gabriel Hiester, Harrisburg, Pa., graduate of the Pennsylvania State College, for many years trustee of the college and president of the State Horticultural Society, had a beautiful farm home near Harrisburg. During the first twenty years in bearing his orchard, of which one-fourth the trees were unprofitable varieties, returned an average of $80 per acre with apples selling at 60 cents to $1 per bushel. Mr. Hiester believed, with a proper selection of varieties and a favorable location, that any well-managed orchard can be made to do much better.]

[Illustration: Dr. J. H. Funk, Boyertown, Pa., graduate of the University of Pennsylvania, 1865, farmers' institute lecturer, former state pomologist, has 50 acres of apples and peaches. Returns from his plantings begun in 1896 are so phenomenal that he is afraid to permit the publication of his profits. It is known, however, that he has sold $5,000 each of peaches and apples in one year.]

COST OF MAINTAINING WORK HORSES

At the Minnesota station, the total cost of feeding and maintaining a farm work horse for one year was estimated to be from $75 to $90, of which about $20 was charged for interest and depreciation. On the basis of 3.3 hours as the length of the working day, the cost per horse per hour was estimated to be 7-1/2 cents. At the Ohio state university, it was found that four horses weighing about 1,400 pounds were chosen to perform 2,185 hours of labor during one year, while under like conditions four horses, weighing about 200 pounds less, worked on an average but 1,641 hours each. For each secular day, therefore, the former worked about 7-1/2 hours, while the latter were employed but five and one-half hours. The cost of food was estimated at $54; cost of shoeing, repairs of harness and stable supplies at $6.50; and the cost of feeding, grooming and cleaning of stables at $23.50, or a total cost of $84 per year. Nothing was charged for interest or depreciation, but the expense of feeding and caring for three colts was included in the estimates given. The annual expense of maintaining a horse was practically the same in both states, but the cost per hour of labor performed was less because of the possibility of employing the horses at productive labor a larger portion of the time. Too much emphasis cannot be placed upon the need of planning a farm organization which will give continuous employment to horses as well as to men in order to realize the most profitable returns. An industrial system that makes it necessary to maintain work animals three days in order to secure one day's work falls far short of an ideal.

CHAPTER XVII

FARM LABOR

The problem of farm labor demands thoughtful and frank consideration. Since work is an essential element in the production of all wealth, it follows that every industry has its labor problem. The adjustment of labor to the production of the various forms of wealth must ever constitute one of the most important problems in any organized society. It is often remarked that the labor problem is the chief difficulty in farming. In a certain sense this is true, since work is a primary element in the production of agricultural as well as all other wealth. It is not true, however, that the problem of labor is more difficult or more intricate than that of other industries. In fact, that problem is less delicate than in some other occupations, because farming is less industrialized.

It is not possible to settle once for all the problem of labor for any occupation, since changing conditions will give rise to new questions or new phases of the old problem. Moreover, the problem of labor on the farm will grow more difficult as farming becomes more specialized and as the methods of production become more complex.

However, the labor problem on the farm is different from that in the manufacturing industries or in trade and transportation. This chapter will not concern itself with an attempt to settle the farm labor problem, but will undertake to state the character of some of the differences between it and other forms of labor and to discuss some of the changes in recent years.

A large proportion of farm work is done by the farm owner, or renter, and his family. There is not much opportunity to profit by the labor of other persons. In 1900 there were in the United States 1,812 industrial establishments each of which employed between 500 and 1,000 persons, while there were 675 establishments each of which had more than one thousand employees. In the same year there were 5,739,657 farms, which employed in the aggregate 4.4 millions of people, not including the owners of the farms. Moreover, over one-half of the 4.4 million persons thus employed were members of the families of the farmer. In other words, aside from members of the family, there was less than one employee to every two farmers. Since a considerable number of farmers employ more than one person, it follows that the majority of farmers employ no help other than members of the family.

In another particular farm labor differs from that of other forms of labor even more widely. There are sociologic as well as economic questions involved. Baldly stated, custom permits, and necessity often requires, the laborer to eat at the same table with the farm owner and in other particulars he mingles intimately with the farmer's family. In all its bearings, this is a very important fact. It constitutes one of the greatest difficulties in the problem of securing suitable farm help. Industrial corporations employ as common laborers largely Italians, Hungarians, Poles and negroes. The English, the Irish, the German, the Swede and the Norwegian have been readily received and assimilated in the American farming communities. The peoples of Eastern and Southern Europe are often criticized because they do not become farm laborers. That they do not is in large part due to the fact that the farm hand is usually a member of the farmer's family. Thus the supply of common labor which is today used by the rest of the industrial world is not open to the farmer.

Farming differs from some other occupations in that it does not ordinarily offer the laborer much opportunity for advancement. The fireman on a railway train becomes the engineer; the brakeman becomes a conductor. There are opportunities in many establishments for the advancement of the industrious and clever. A man may enter their service with the hope of being able to marry and support a family. On the other hand, all our land laws are based upon the idea that each farm should be of sufficient size to support only one family. Where it does support two families, the relation is usually that of landlord and tenant. The farm laborer, therefore, must look upon his employment as more or less temporary. The young man who intends to become a farmer will find employment upon the farm a desirable if not essential preparation for his future occupation.

The introduction of farm machinery has had the effect of increasing the price of farm labor while at the same time decreasing the amount of labor needed. The reason is that the introduction, not alone of farm machinery, but all forms of machinery, has made man's labor much more efficient than formerly. Farm wages have doubled since the introduction of horse-drawn machinery. The labor income in the different sections of the United States is influenced by the extent and efficiency with which machinery is used. The relation of labor income to the use of horse power is shown by the following table taken from a recent census:

INFLUENCE OF FARM MACHINERY AS SHOWN BY THE RELATION OF LABOR INCOME TO HORSES AND MULES.

Divisions of the United States	Labor Income	Number of horses and mules to 1,000 persons in agriculture
North Atlantic	$299	1,655
South Atlantic	163	808
North Central	402	3,036
South Central	211	1,603
Western	510	5,476
United States	$288	2,105

In one of the states of the South Atlantic division the average price of farm labor, without board, was $12 per month, while ir one of the states of the western division the price on the same date was $31. Why? Because in the latter case a man's labor was more productive. In the South Atlantic division, in producing the chief crops cotton and maize, a man uses one mule in preparing and cultivating the soil. In the western division plowing and harrowing with six-horse teams is common and nine-horse teams are not unusual. The cotton picker in one day will be able to gather not to exceed 300 pounds of seed cotton, worth not more than $15. The western wheat will be harvested by a machine drawn by 28 horses. In the same time four men with this outfit will cut and thresh 700 bushels of wheat, worth $500.

When the threshing machine was first introduced in Ohio, it was stubbornly opposed by all farm laborers. "They claimed it," says Bateman, "as a right to thresh with a flail, and regarded the introduction cf machinery to effect the same object in a few days which would require their individual exertion during the whole winter, not only as an invasion of a time-honored custom, but as absolutely depriving them of the means of obtaining an honest livelihood. At a later date, when a reaper had been introduced into a field of ripe wheat as a matter of experiment only, every one of the harvest hands deliberately marched out of the field and told the proprietor that he might secure his crop as best he could, that the threshing machine had deprived them of their regular winter work twenty years ago and now the reaper would deprive them of the pittance they otherwise could earn during harvest." How short-sighted they were! No class gained so much from the introduction of labor-saving machinery as did those who did the labor. The reason for the increase in well-being, the reason society enjoys luxuries and comforts beyond the fondest dreams of former generations, is due to the fact that the labor of each man has been made so much more effective through these labor-saving devices. The humblest citizen shares in this improvement. Not all share alike and not all share equitably, but each generation sees its members sharing more equitably than those of any generation which preceded it.

The proposition is an extremely simple one. If a man produces just enough food for himself and family, he will have nothing for clothing, shelter, or education. If, however, a man produces four times as much food as he and his family consume, he may exchange one-fourth for shelter, one-fourth for clothing and have remaining a fourth for education, and recreation or savings. This is only another way of saying that the greater the amount of any useful commodity produced by a single day's labor the larger will be the laborer's income or wages.

Although the increase in intensive agriculture and the diversification in farming tend to increase the need of farm laborers, the introduction of farm machinery has much more than offset this demand. The tendency of farm laborers to become farm tenants; or, to state it in other words, the tendency of landowners to rent their land rather than to continue to operate it themselves, is not without its influence upon the labor problem.

The invention and introduction of farm machinery has accentuated the difficulty of keeping the farm laborer continuously employed. The decrease in the demand for farm labor and the increasing lack of uniformity in the amount required have caused a gradual depletion of the smaller villages and hamlets which were a source of labor supply during harvest and other busy seasons.

The problem of keeping labor continuously employed has always been a difficult one on the farm, because of the change of seasons and because of the variations in the weather from day to day. There is a wide difference between those industries which are carried on within doors and farming, which is subject to the caprices of the weather. Natural causes produce tremendous variations in the return for labor. For example, in 1901 there were produced in the aggregate 3,006 million bushels of wheat, maize and oats, while in 1902 there were harvested 4,180 million bushels. Here is an increase of over a thousand million bushels. The same farmers tilled the same soil in the same way as far as natural causes would allow, and yet there was a difference in result amounting to 39 per cent. A variation of one hundred million bushels of wheat from year to year, due to climatic conditions solely, is not at all unusual.

The manufacturer also has far greater control of his labor. When it rains, he has a roof over his workmen, and hence the work is not interrupted. When it grows dark, he turns on the light and the work continues. If it gets cold, he lights the fire and still the work continues comfortably. It is not so in agriculture. There is a great variation in the working efficiency of men employed in farming. In a certain locality there were twenty-one days of rain in the thirty-one days of May. The next year between June 5 and September 5 in the same locality there was not half an inch of rainfall at any one time.

What is true of labor is also true of machinery. The farmer must purchase machinery which he can use only a few days in the year, while the manufacturer, for the most part, employs his machinery continuously, sometimes day and night. While natural causes prevent the farmer from using the same business methods, or from being able to calculate his profits with the same precision as is possible by those following manufacturing and mercantile pursuits, it is nevertheless important that farming should be planned to avoid, as far as possible, the influence of natural causes. Certain kinds of farming are less dependent upon natural causes than others. Wisdom and foresight can do much to avoid, in all farming, untoward influences. The clever farmer seldom complains about the weather.

Farm machinery has made unnecessary, and hence unprofitable, some of the labor at which children were formerly employed. In the not distant past many, perhaps most farmers, owed their prosperity in large measure to the labor of their children. A large family, especially of boys, was a valuable asset. Even a generation ago conditions were not far different, and two generations ago were quite the same as those described by Homer:

"Another field rose high with waving grain: With bended sickles stand the reaper train: Here, stretch'd in ranks, the level'd swaths are found; Sheaves heaped on sheaves here thicken up the ground. With sweeping stroke the mowers strow the lands; The gath'rers follow, and collect in bands: And last the children, in whose arms are borne (Too short to gripe them) the brown sheaves of corn. The rustic monarch of the field descries, With silent glee, the heaps around him rise. A ready banquet on the turf is laid Beneath an ample oak's expanded shade. The victim ox the sturdy youth prepare: The reapers due repast, the women's care."

There is also another reason why the age of the employed has been raised. It is due to the growth of higher education. Where formerly the farmer's children between the ages of twelve and twenty-one did most of the farm work, now many of them at the same age are attending schools and colleges. The sons of a man, who a generation ago found no opportunity to get beyond the district school, graduate from high school and college, and thus spend most of their time in study until they are past twenty-one years of age.

Labor unions have doubtless caused a scarcity of farm labor by increasing the proportion of the created wealth which goes to the man who labors without capital. When a man can obtain fifty cents an hour for laying brick, he does not wish to work in the hay field at twenty cents an hour, even though the difference in the cost of living may in great measure offset the difference in wages.

There is a growing tendency to perform work by what is called contract labor. Thus a person may agree to weed and hoe sugar beets at a certain rate per acre. He, in turn, employs a force of cheap laborers which he sends from farm to farm to do this work. The harvesting of fruits and garden crops is not infrequently done in some such manner. In one instance a contractor of laborers of foreign birth has been furnishing them for all kinds of farm work. He keeps 20 to 40 of these laborers on a small farm, furnishing them a dwelling and selling them food supplies. Farmers telephone for help when in need. The contractor receives $1.65 for a day's work and pays the laborer $1.50.

It appears from the preceding considerations that there are open to every farmer at least three methods of increasing the efficiency of farm labor. He may make every day's labor more efficient by use of labor-saving machinery and the employment of it in the most efficient manner; as, for example, using three 1,500-pound horses to his farm machinery instead of a pair of 1,200-

pound horses. He may modify the character of his farming in order that profitable labor will be more continuous. He may modify the method of employing labor; as, for example, by introducing the system of contracting labor for specific purposes where feasible.

Increase in the price of farm labor is not an evil. It is an indication that labor applied to agriculture is becoming more productive and hence more profitable. Since more than one-half the labor of the farm is done by the owner and his family, the farmer is benefited through the rise in price of farm wages. The more that labor can be made to earn upon the farm, the better it will be not only for the farm owner but for society in general.

CHAPTER XVIII

SHIPPING

The means of facile transportation and the machinery of trade are the need and the development of a complex civilization. The importance of these useful adjuncts of everyday life is indicated by the fact that about one-fourth of all the people engaged in gainful occupations in civilized communities are employed in them. Nevertheless the expense of transportation and trade constitutes a tax upon the consumer which it is the aim of modern methods to reduce to the lowest limits. Recent investigations indicate that for every thirteen dollars the consumer expends for farm products the producers receive six dollars. In some directions most remarkable results have been accomplished. A recent quotation on wheat per bushel was as follows: Chicago, $0.93; Antwerp, $1.04; London, $1.06; Hamburg, $1.07. Eleven to 14 cents per bushel represents the cost of haul and commissions between Chicago and the European cities named. Methods of handling have been so perfected that from the time the western farmer places the bundle of wheat at the mouth of the threshing machine the grain literally flows through the channels of trade until it reaches the flour sack. On an average the English miller pays about 20 cents a bushel more for wheat than the American farmer receives for it.

The cost of distributing many other farm products is greater, although the range of distribution is much less. The cost of haulage and selling potatoes is from 25 to 50% of the retail price, while with hay it is still higher. The cost of distributing all forms of truck and market garden produce is high and often wasteful. Many attempts have been made to eliminate a part of this cost as well as to better the conditions of the supplies when they reach the consumer. While many individuals have been quite successful in dealing directly with the consumer, little has thus far been accomplished that affects general trade conditions. Great improvements have been made in methods of transportation and methods of preservation. Cold storage and canned goods have been the direction in which progress has been notable.

WASTEFUL METHODS OF DISTRIBUTION

Owing to customs and traditions there is frequently a great waste of effort in some of the methods of trade. The meat trade of France is an excellent illustration. Certain sections of France make a specialty of rearing cattle. At a suitable age these animals are purchased by other farmers who fatten them. Many of the small towns maintain market places at which fairs are held to facilitate these negotiations. Frequently there is a shipment from one region to another, which is conducted by a middleman. When fattened the steers are collected by a stock buyer, who may ship them to La Villette, the live stock market of Paris. Here they are placed on sale through commission men. There are the usual charges for yardage and food. After being sold the animals are driven to the slaughterhouses. The carcasses are then taken by wagon to the great market of Paris located near the center of the city. Here the retail vender of meats comes, makes his purchase, reloads the meat, which may have been unloaded less than an hour before, carries it to his shop, where the consumer seeks it. The number of people concerned and the amount of hand labor have been excessive.

Nor is the American system without its faults. The Iowa or Illinois farmer fattens cattle that may have been reared in Montana or Texas. After the stock buyer, the commission man and the stock yard company have each taken his toll, the packer ships the carcasses back to the very region where the animals were fattened, when the stockman may purchase it of the local vender of meats. The facilities and perfection with which these many transactions are accomplished is one of the wonderful sights of our country. Nevertheless the producer of meat products may well consider whether some more economical system of distribution may not be devised.

SHIPMENTS: SOURCES OF INFORMATION

All railroad rates are now carefully supervised by the federal government and are open to the inspection of the public. Such information as is ordinarily needed may be obtained from the local station agent, who is always glad to be of service to patrons of his road. If information of a special character is required, it may be obtained by addressing the division freight agent of the railroad in the region under consideration. The name of this officer is to be found in the circulars and upon the posters of the railroad.

In addition to the freight facilities offered by any individual railroad, there are what are known as fast freight lines. These agencies enable through and prompt shipment from inland points in our own country to inland points in another. An individual railroad may operate in connection with several such agencies. A certain railroad, for example, is combined with nine fast freight lines. Freight agents of local roads in the principal towns usually represent the fast freight lines and are prepared to transact business.

In seaport cities there are firms styling themselves foreign freight contractors, outward freight agents, steamship agents, or ship brokers. These firms are prepared to quote prices on shipments to any part of the world on either regular or tramp ships. They will give freely to intending shippers full information concerning methods and conditions of shipment. There is nothing mysterious about the business of shipping farm products. The necessary details may be acquired by inquiry in the channels indicated and by a little study of the data, which will be cheerfully furnished.

RAILROAD RATES

A great many factors are involved in determining the rate which is charged for transporting different products. In a certain sense it is doubtless true that the rate charged is based upon what the traffic will bear. The purpose here, however, is to state some of the customs which exist rather than to discuss the philosophy or justice of them.

The rate may vary with the value of the product, without any regard to the cost of the haul. Suppose the cost of shipping a ten-gallon can of fresh milk between two points to be 32 cents, the cost of shipping a similar can of cream may be 50 cents. The cost of shipping a carload of hay is less than a carload of wheat.

In some instances, zones or belts have been recognized, the rate from all towns within each zone being the same for a given product. Certain railroads centering in New York recognize four zones for the shipment of milk and cream, as follows:

Zone A--First 40 miles. Zone B--Between 40 and 100 miles. Zone C--Between 100 and 190 miles. Zone D--Beyond 190 miles.

It will be noticed that the size of these zones varies and may be the subject of adjustment between railroads and shippers.

While less understood by the public, railroads recognize zones or, more properly, groups of towns in making rates to them instead of from them, as in the instance above mentioned. It is possible to change the rate on a product to a given town by classifying it in another group. The rate on bran and other stock foods from central western points to certain towns in New York state has been the same as that charged to Boston, Mass., while other towns in New York not far removed have taken a lower rate.

Differential rates are recognized to be legitimate. Railroads are allowed to charge a less rate for wheat intended for export than that intended for local consumption. There has sometimes been a wide difference between the freight rate on wheat between Kansas City and Galveston, Texas, depending upon whether the wheat was to be exported or intended for domestic use.

In certain sections and for certain products the railroad rate varies with the season, because of difference in competition. The railroad rate between Chicago and New York on grain is higher while the navigation of the Great Lakes is suspended. As an illustration of the cheapness of transportation by water, it is stated that sometimes it is cheaper to ship wheat from Chicago to Buffalo by boat than to store it in a grain elevator for an equal period of time.

Products may sometimes be sent by baggage to greater advantage than by express, special arrangements for which are generally required.

FACILITIES FOR FREIGHT TRANSPORTATION

American railway facilities are, perhaps, unrivaled among the nations of the world, but the United States is still behind other nations in the matter of means of local transportation, in which good roads is only a part of the problem. In France, the so-called messagers are a common feature of local traffic. Thus in the Department of Touraine there are 246 towns each having from one to four messagers, who with their great two-wheel carts, each with single draft horse, make one or two trips to Tours each week. The messagers carry freight both ways precisely in the same capacity as railroads do. While the railroads are fairly abundant these local agencies continue to thrive because delivery can be made directly to the consignee and delivery at the exact time and place is more certain. The enormous loads conveyed in these two-wheel carts by one horse is an element in this system to which the good roads of France now contribute. In 1799, France had constructed 25,000 miles of roadway. Since that time, over 300,000 miles of roadway have been completed and about 30,000 miles of railway have been constructed--ten miles of roadway for each mile of steam railway. The good roads of France are of comparatively recent origin, contributing materially to the improvement in well-being which has taken place during the same period.

CHAPTER XIX

MARKETING

Without stopping to inquire the reasons, it may be recalled that there are two rather distinct forms of trade, wholesale and retail. The wholesale trade is conducted by three classes of persons: dealers or merchants, commission men, and brokers. The dealer is one who buys the goods outright and takes his own risk on making a favorable sale to the retailer. The commission man is one who receives the goods, sells them at such price as he may be able to obtain and remits to the seller the amount obtained less expenses and his commission. The broker is a man who effects a sale without coming in contact in any way with the materials sold. A cheese broker, for example, receives instruction from different factories to sell for them a certain quantity of cheese of a given kind and quality each week or month as the case may be. At the same time he receives from grocery stores which retail cheese orders for various amounts, kinds and quality of cheeses. With this information at hand, he directs the various factories intrusting their business to him to ship the kind, quantity, and quality of cheese required by his several customers. For such service he receives a brokerage, which is less than that charged by a commission man because he is not required to handle or store the material.

Since the different farm products are purchased by different classes of retailers, and since their handling and sale require different facilities and special knowledge, there have arisen in the great centers of trade different kinds of markets, each having its particular facilities for the handling, care and sale, and each conducted by commission men or brokers with a special knowledge of the trade. Furthermore, certain cities have become, on account of their favorable position--to mention but one reason--headquarters for certain products or groups of products. Thus Petersburg, Virginia, has the principal wholesale market for peanuts. Elgin, Illinois, has been noted for its butter market. St. Louis is the leading mart for mules.

In a general way, the following five more or less distinct and important classes of markets for farm products may be recognized: Grain, Live Stock, Produce, Cotton and Tobacco.

METHODS OF TRADE

The brokers or commission men doing business in any one of these markets usually form an association called a board of trade, chamber of commerce or similar title for the purpose of assisting "each other in the pursuit of common ends." The result has been uniformity of methods and charges; but above all in importance, perhaps, has been the definition of classes and grades of the products placed on sale. The tendency is for the associations in the different cities to adopt uniform rules for the grading of products, so that No. 2 red winter wheat may mean the same thing in Toledo and New York; that the quotation on prime beef may refer to the same quality of cattle in Pittsburgh as it does in Chicago; and that No. 1 Timothy hay in Baltimore and St. Louis may be alike. While the tendency is towards uniformity, much yet remains to be accomplished. The shipper must be on his guard lest he suffer loss through the variations in the classification or variations in their interpretations on the different markets.

There has grown up around these markets some agency which stands as a disinterested party between seller and buyer impartially determining the weight and in some cases the quality of the object under negotiation. The State of Illinois employs agents who inspect all cars of grain consigned to the Chicago market. These inspectors determine the kind, grade and weight of the grain in each car. The car is then delivered under seal to the purchaser. If either seller or buyer is dissatisfied with the inspector's decision he may, by complying with certain regulations, have this decision reviewed by a higher authority. The decision of this higher authority is final and must be accepted by both parties. Brokers selling grain in carload lots ship the cars subject to the weight and grade as determined by the inspector at Chicago. Grain of a specific grade may thus be bought in Chicago or other great grain markets with almost perfect security as to weight and quality by persons living in any part of this or any other country. At Elgin the quality of butter is determined by a committee appointed by the Board of Trade from its own members. In the live stock markets, the stock yards company, in addition to furnishing yards, shelter, food and water, acts as agent between seller and buyer in determining the weight of the animals. The purchaser or his agent must determine for himself the quality of the animals he buys.

GRAIN MARKETS

The Chicago and St. Paul Boards of Trade and the New York Produce Exchange are the three great agencies for dealing in grain in the United States. Buffalo, Duluth, Baltimore and Philadelphia are also important markets. Adjuncts to these markets are the great terminal elevators capable of holding almost indefinitely enormous quantities of wheat and other grain. On the Pacific Coast all the wheat is handled in the bags, as is the custom in the other markets of the world. Canada and the United States alone have recognized the principle that wheat and other grains will run like water, which has been a prime factor in their competition with other nations.

Country elevators charge two cents a bushel for storage during the first 15 days and 1/2 cent for each additional 15 days. The charge for storage at terminal elevators for the first 15 days is 3/4 cent. The farmer may thus store his wheat in an elevator in place of his farm if he chooses so to do, although the wheat he thus puts in storage may have been made into flour and consumed before he sells it. This may be looked upon as a sort of intermediary step between storing wheat in one's own granary and dealing in futures.

The country shipper pays 1/2 cent a bushel commission for the sale of wheat. There is also a charge for inspection and insurance, and, in case there is an advance payment, for interest. After five days there are storage charges. This has given rise to the expression, gilt edge, regular and short receipts, depending upon the length of time there remains before storage charges must be paid. Every market has a grade known as contract grade, meaning the quality that must be furnished when wheat or other grain is sold without specifying the grade. In Chicago No. 2 red winter wheat is the contract grade. Where grain is sold or purchased by a broker, the brokerage is usually 1/8 cent per bushel.

HAY MARKETS

At least twenty cities have adopted the rules of the National hay association as to classes and grades of hay and straw. The southern states constitute an important market for the hay of the north central states, while Boston, New York and the mining towns of Pennsylvania are important markets for the northeastern states. The size of bale varies from 75 to 200 pounds. Small bales of 100 pounds each are preferred in Baltimore, medium bales of 110 to 140 pounds in Philadelphia, while New York and Boston usually deal in the larger bales. The commission charges vary from 50 cents to $1 per car. In New York, $1 pays all charges. At Chicago, $3 per car has been charged for the inspection, divided equally between seller and buyer.

PRODUCE MARKETS

Every town of any consequence has its produce market. The South Water street district in Chicago and the West Washington street market in New York are noted for their extent and variety. There are also many special markets for certain classes of produce. Thus Elgin, Chicago and New York have butter exchanges. Wisconsin, Utica, Watertown and Cuba (New York) maintain exchanges where cheese is placed on sale each week during the manufacturing season. There is also a board of trade for cheese in New York City. The prices quoted upon these exchanges are made the basis of many transactions between buyer and seller, who never enter these markets. Not only do buyers and sellers agree to abide by the quotations of one or the other of these markets, but the quotations are also used as a basis of settlement for milk furnished the creamery or factory. These agencies are thus impartial arbiters in countless financial transactions.

The rate of commission varies in different markets and for different products. Generally, however, produce is handled on a 5% basis, but for individual products which are especially bulky and difficult to handle, such as cabbage, 10% may be charged. In some cases commission is by quantity instead of on a percentage basis. Thus for potatoes the commission is sometimes 10% and in other cases 4 or 5 cents a bushel.

LIVE STOCK MARKETS

While poultry and game, as well as the carcasses of the smaller animals, may be handled through the produce markets, the large animals require separate facilities. The United States is noted for its large live stock markets and for the perfection and size of the packing houses which have grown up about them. The most famous example of these combined agencies is to be found at Chicago, but important live stock markets are also maintained at St. Louis, Kansas City, Omaha, Pittsburgh, Buffalo and more recently Fort Worth, Texas. The commission charges vary from 50 cents to $1 per head for cattle and from 10 to 25 cents per head for calves, sheep and hogs. In some markets, the commission on hogs is 2% of the gross returns. When located within 150 miles of a central market, it is customary to allow 50 cents per hundred pounds for cattle and 40 cents for hogs to cover shrinkage, and cost of freight, yardage, food, bedding and commission. It is possible for an owner to sell his own live stock in these yards, but the commission man, because of his superior knowledge of existing trade conditions, is almost universally employed. Firms which handle cattle, sheep and hogs seldom sell horses. Although handled by different commission firms, important horse markets are maintained at Chicago and Buffalo immediately adjacent to the market for meat animals. In New York the horse markets are in a different section of the city, that for draft and common work horses on one street, while the American Horse Exchange, located at another point, handles high-class light horses. The usual custom is to sell horses at auction, although they may be purchased at private treaty. In whatever manner purchased, it is essential to understand precisely the character of the guarantee.

COTTON, WOOL AND TOBACCO MARKETS

Because of their higher value per pound and the ease with which they can be stored, cotton, wool and tobacco are dealt in somewhat differently than other farm products. The two great cotton exchanges are located at New Orleans and New York, the quotations on these markets controlling the financial transactions in cotton throughout the world. The principal wool markets are Boston, New York, Philadelphia and St. Louis. The principal tobacco markets are at Richmond and Danville, Va., Durham, N. C., and Louisville, Ky.

[Illustration: Mr. C. W. Wald, graduate of the Ohio State University, farmer, formerly assistant horticulturist of the New Hampshire and Ohio Experiment Stations, is shown above in one of the New Carlisle (Ohio) greenhouses, in which three crops of lettuce occur in one bed. One crop is ready to cut, another planted and a third in pots between the other plants, to be planted in another bed when large enough. The net returns from a quarter of an acre under glass has been greater than from 160 acres devoted to general farm crops.]

[Illustration: C. W. Zuck & Sons, Erie, Pa. One son was a student in agriculture at the Pennsylvania State College. Father and three sons, beginning six years ago with a run-down farm of 55 acres, have built an acre of glass and a heating plant of 260 horsepower. During the period they have spent $5,000 on the place and at the end of season they will have very nearly cleared their improvements. "Tell the youthful readers of your book to get as much education as possible and then go in partnership with their fathers or brothers. If they do, success will be theirs."]

The country shipper or the young farmer wishing to place his products in the ordinary channels of trade must consider and determine among other things the following: What cities have favorable markets for his products; choose some commission man or broker to handle them; calculate the expenses for freight, commission and other customary items; familiarize himself with the rules for grading his products in the market or markets under consideration; and determine what agency there may be for protecting him as to the weight and quality when sales are effected. Whenever practicable, a visit to the market in question and a personal study of the conditions under which selling is done will be wise. Having done so, and perhaps having made a number of sales through these usual channels of trade, he will be in a position to consider whether he may organize to advantage some more direct method of getting his products to the consumer.

CHAPTER XX

LAWS AFFECTING LAND AND LABOR

Thus far property has been treated as invested capital upon which interest must be charged in determining the labor income. Labor, likewise, has been considered principally in its effect upon profits. Society has thrown around the transfer of property and the use of labor certain restraints for the protection of all individuals.

Through the ages certain procedures have become fixed by custom. These legal practices are largely the inheritance of old Roman law and are usually known as common law. Various legislative bodies having jurisdiction enact from time to time other laws. This body of enacted law is called statute law and is much more variable than common law. In the briefest possible manner it is the purpose here to state a few of the principles and applications of the law, chiefly the common law, as it affects the farmer in acquiring or disposing of his property and in his dealings with labor.

PROPERTY

Property may be defined as anything which is a subject of ownership. It possesses the characteristics of being acquired, held, sold, willed or inherited and is of two kinds: (1) Real property, real estate or realty; (2) chattels or personal property. These two kinds of property are subject to quite distinct legal practices. In general, real estate consists of land, things attached to it, such as trees, buildings, fences and certain rights and profits arising out of or annexed to the land. The term land as ordinarily used includes all these things, so that when land is said to be worth so much an acre it includes all fixtures. Ponds and streams are, under this definition, land. The land not only has surface dimensions, but extends upward indefinitely and down to the center of the earth, and hence includes a right to ores, coal, oil, gas or other materials whatsoever.

An article may, however, be real property or personal property depending upon circumstances. Thus a tree growing on the land is real property, but when cut into cord wood becomes personal property. New fence posts ready for use are personal property. When set in the ground they become real estate. Just what goes with a farm or what are fixtures is frequently a subject for legal determination.

FIXTURES

The general rule is that "fixtures are any chattels which have become substantially and permanently annexed to the land or to buildings or other things which are clearly a part of the land."[D] The annexation may, however, be purely theoretical, since the keys to the house or barn, which may be in the owner's pocket, are real estate. One rule concerning fixtures is that they must be so annexed that they cannot be severed without injuring the freehold. The intention of the party making the annexation also often determines, since if the article is annexed with the intention of making it permanent, it then becomes a part of the land. Among the things held to be fixtures, and therefore a part of the land, are: (1) All buildings and everything which is a part of any building, such as doors, blinds, keys, etc.; (2) fence materials which have been once used and are piled up to be used again are a part of the land, but new fence material not yet used is personal property. (3) Growing crops are real property. They go to the purchaser of the land unless specially reserved in the deed. A verbal agreement is not sufficient. (4) Trees, if blown down or cut down and still lying where they fell, are real property; if cut or corded up for sale they become personal property. (5) All manure made on the farm is real estate and passes with the land. (6) All the ordinary portable machines and tools are considered personal property, but certain machines held to be of permanent use upon the land are real estate. Among the things which courts have held to go with the land are cotton gins, copper kettles encased in brick and mortar for cooking food for hogs, cider mills, pumps, water pipes bringing water from distant springs. In general, motive power machinery and the shafting go with the land, but the machinery impelled may or may not, depending upon the way it is annexed. (7) If stones have been quarried for the purpose of using upon the farm, they go with the farm, but if quarried for sale they are personal property.

CONTRACTS

The difference between personal property and real property may be indicated by considering the essential features of a contract. A contract is an agreement between two or more persons. The foundation rule concerning a contract is that every man must fulfill every agreement he makes. An ethical practice grows out of this legal rule which, if strictly adhered to, will save much embarrassment, viz., make but few promises and always keep your engagements.

There are seven requirements generally necessary to a valid contract. (1) Possibility. The thing to be done must be possible. (2) Legality. It must not be forbidden by law. (3) Proper parties. The parties to a contract must be competent. Contracts with idiots or drunken persons are not binding. Some contracts with minors are not binding, although contracts for the necessities of life are. (4) Mutual assent. A proposition not assented to by both parties is not binding on either. (5) Valid consideration. A man is not regarded as injured by the breaking of a promise for which he has paid, or is to pay, nothing. (6) Fraud or deceit. A contract obtained by fraud is void as against the party using the fraud, but may be enforced by the innocent party if he sees fit. (7) Written contracts. Here comes the most important difference between real and personal property. Real property can only be conveyed by a written instrument, properly executed and recorded, while personal property passes by mere possession. Contracts relating to the sale of real estate are not binding unless in writing, while verbal contracts are sufficient for personal property if accompanied by payment of a part of the purchase price or the acceptance of the goods. For amounts under $50 verbal agreement in itself is binding.

TRANSFER OF REAL ESTATE

The purchaser should require of the seller evidence that the title to the land is straight and clear; if not, exactly what the defects are. This is done through an abstract of title, which should be prepared by a competent lawyer. This is not an official document, and its value depends largely upon the ability and watchfulness of the party making the abstract. Ownership of land is conveyed by means of a deed. A deed is an instrument conveying at least a life interest in the land. Care should be taken that the deed contains the essential parts and that it is properly executed.

DEEDS

Deeds are of two kinds: Quit claim deeds, which convey all the rights, title and interest which the seller has in the land, but does not warrant the title; and warranty deeds, which, in addition to what a quit claim does, contain covenants which agree that the seller and his heirs, etc., shall warrant and defend the title to the purchaser against the lawful claims of all persons.

THE REQUISITES OF A DEED

The requisites of a deed are: The parties to the deed, the consideration, the description; and with a warranty deed, the covenants. The seller must be of full age, sound mind and if married his wife should always join in the deed. Her name should appear following his at the beginning of the instrument. She should sign and acknowledge the deed, and the certificate of acknowledgment should state that she is the wife of the seller. If the seller is a married woman, her husband does not need to join in the sale of her own property. It is customary to state the consideration upon which the deed is given, but this is not necessary, nor will a false statement as to the amount paid invalidate the deed.

The description of the land conveyed should be as minute and careful as possible, and preferably in the exact language of former deeds. In case former description is in error, it should be referred to and correct description given. Where land is conveyed by metes and bounds, this description governs, although it may not convey the number of acres of land stated. In describing boundaries the location of monuments takes precedence of distances mentioned.

EXECUTION OF THE DEED

A deed must be signed, witnessed, acknowledged, delivered and recorded. In some states deeds must be sealed, but in other states the law has dispensed with this formality. Witnesses to deeds are not required in all states. Some states require one, but usually two witnesses are required. The parties signing the deed are required to appear before an official designated by statute, usually any magistrate, justice or notary public, and acknowledge the same to be his or her free act and deed.

A deed has no effect until delivered, and should be immediately recorded by the purchaser. Generally an unrecorded deed is not good as against a subsequent purchaser in good faith. It is well to note that the laws relating to the transfer of land are those of the place where the land lies and not necessarily those of the place where the deed is made.

METHOD OF LAYING OUT PUBLIC LANDS

The public lands of the United States are, whenever practicable, laid out into townships each six miles square, "as near as may be," whose sides run due north and south and east and west. The townships are laid off north and south of a base line which is a parallel of latitude, and are numbered north and south from the base line: Thus, T. 3 S., means Township No. 3 south from the base line. Each row of townships running north and south is called a range, and is numbered east or west of the principal meridian: Thus, R. 2 E., means Range 2 east of the given meridian.

The townships are then laid off into sections or square miles of 640 acres, "as near as may be," and these are numbered, beginning always at the northeast section, as shown in the accompanying diagram.

```
N +----------------------------+ | 6 | 5 | 4 | 3 | 2 | 1 | |----+----+----+----+----+----|
| 7 | 8 | 9 | 10 | 11 | 12 | |----+----+----+----+----+----  W | 18 | 17 | 16 | 15 |
14 | 13 | E |----+----+----+----+----+----| | 19 | 20 | 21 | 22 | 23 | 24 | |----+----
+----+----+----+----| | 30 | 29 | 28 | 27 | 26 | 25 | |---+----+----+----+----+----| |
31 | 32 | 33 | 34 | 35 | 36 | +----------------------------+ S
```

Each quarter section is referred to as the northeast or southwest quarter of the section, and each forty acres as the northwest or southeast quarter of a particular quarter. For example, an eighty-acre field may be referred to as the west half of the southwest quarter of Section 3, Township 5 North, Range 3, west of ----. Base line and meridian, or in some cases merely the meridian is mentioned.

The curvature of the earth's surface makes it impossible for the sides of townships to be truly north and south and at the same time six miles square. The excesses and the deficiencies due to the convergency of meridians and the curvature of the earth are by law added to or deducted from the western and northern ranges of sections and half sections of the townships. While the above has been the rule in laying out public lands for more than a century, there are many exceptions, due to many causes.

In the older settled sections the land was laid out in lots, often in a very irregular manner, although in some cases within a given tract the area was more or less regular. In these cases, the land must be described minutely and carefully by metes and bounds. In some of the southern and western states, also, where there were Spanish grants, much irregularity in the surveys exists. Over much of the north Central states this rectangular system of laying out lands obtains and has worked well in most respects.

THE LANDLORD AND TENANT

Leases of real estate follow the same procedure as deeds, except that a verbal lease, if for a term of not to exceed one year, is valid in most states. A written lease should be carefully drawn, because, according to common law, there are few things implied in a lease that are not stated. Definite statement concerning repairs and insurance is desirable. A tenant should also acquaint himself with the law of the state concerning the surrender of the farm upon the expiration of his term.

It is the duty of the tenant not only to guard the property, but to conduct the farm in a husbandlike manner. Unless otherwise stated in the contract, the tenant must pursue those methods of husbandry which are customary in the vicinity.

THE RELATION OF THE FARMER TO HIS WORKMEN

The requirements of a valid contract, as previously stated, control most of the relations which the employer has with his employees. Contracts for labor, unless for more than one year, need not be in writing. If, however, the service to be rendered is unusual, the agreement should be reduced to writing, because, in the absence of specific agreement, the law assumes that customary service and wages are implied.

Like all other employers of labor the farmer is under obligation to protect his workman from injury. He must not subject them to unusual and unreasonable risks. He must hire workmen suited to the employment. For example, if he employs a young boy to drive a fractious horse, he would be liable for any injury that might occur. In like manner, he must exercise proper care concerning the safety of the machinery placed in the hands of his workmen. He must keep his premises in a safe condition and must not expose his workmen to risks not incident to the employment for which they are hired.

The farmer is liable in damages for the acts of his workmen which are within the scope of their employment, although the authority may not have been expressly conferred. "He who acts by another acts himself." In case one is sued for the acts of his employee, the burden is upon him to prove that the act of the workman was without authority, expressed or implied.

[D] Haigh's "Manual of Law," p. 69.

CHAPTER XXI

RURAL LEGISLATION

Various laws have been enacted by federal and state legislatures for the better protection of producer and consumer. Much of this legislation affects in a very special way the interests of the farmer. Not infrequently, in fact, generally, the state department of agriculture has more or less direct jurisdiction over their enforcement. State departments of agriculture usually publish a collection of the laws of this character. These laws vary greatly in the different states and only the most general outline, as they affect the interests of the farmer, can be given here. Persons can inform themselves as to the details as enforced in a given state by applying to the state secretary of agriculture.

A number of these acts affect interstate commerce, concerning which the United States Constitution says: "No state shall, without the consent of the Congress, lay any impost or duties on imports or exports, except what may be absolutely necessary for executing its inspection laws." By a series of judicial decisions it has been determined that a State has a right to enforce laws affecting interstate commerce when traffic in the articles thus modified or prohibited affects the public welfare. When it is necessary to have a police regulation to prevent fraud in the traffic of an article or for the purpose of guarding the public health or morals, police laws, so called, may be enacted and enforced. Around this general question there has waged a bitter controversy which has occupied some of the best legal minds and is one involving some difficulty.

FERTILIZER CONTROL

One of the first of the "control" measures to be enacted, and the one which has been most universally adopted by the several states, is the law requiring the manufacturer and dealer in commercial fertilizers to guarantee the percentage of the so-called essential fertilizing elements--nitrogen, phosphorus and potassium--contained in each bag of fertilizer offered for sale. Subsequent control laws have been modeled more or less closely after this law. Hence a description of the operation and execution of it will serve for all.

The execution of this law is usually under the immediate supervision of the state secretary of agriculture, while the necessary chemical analyses are made by the state experiment station. In some states the enforcement of the law is in charge of the state experiment station, while in others the state department of agriculture has its own laboratories or employs a private chemist. It is, however, becoming a more and more settled policy to place all police regulations in charge of the state department of agriculture, while at the same time the chemical analyses and other scientific and technological inquiries are made at the state experiment station.

In order to facilitate the taking of samples and in order to raise funds for the execution of the law, the manufacturer is required to take out a license and to make a statement of the brands of fertilizers which he will place upon the market in the given state during the given season.

During the spring and fall season agents traverse the state and sample the bags of fertilizers as found on sale by local merchants. The samples are sent by number under seal to the designated chemist, while at the same time the agent transmits to the state officer in charge of the enforcement of the law the necessary information concerning these samples. Upon the receipt of the analysis made by the chemist, who has had no knowledge of the origin of the sample, the state officer compares them with the guarantee of the manufacturer, and if he finds it necessary enters legal complaint. While these laws have been in force for many years in some states and in many states for some years, prosecution has seldom been found necessary. The honest manufacturer is protected from dishonest competition, and the dishonest manufacturer, if there be such, cannot afford the publicity which noncompliance with the law would entail.

It has been customary to publish, with the results of analysis, also an estimate of the commercial value per ton of each brand of fertilizer. This estimated commercial value is obtained by multiplying the pounds of each element or combinations of the element in a ton by a value per pound. To the value of the fertilizer thus obtained is added something for cost of mixing, bagging and freight, and something for profit. The price per pound given to each element or combinations of the elements is based upon the commercial value of the element when purchased in raw materials. The price for each year is usually determined by a conference of those in control of the execution of the law in the several states for certain groups of states. As a matter of fact, the price varies little from year to year.

The published figures, therefore, constitute a table of comparative commercial values as determined by the most expert knowledge. While not constituting a statement of absolute commercial value for any given locality, they do enable the purchaser to determine whether the price quoted on a given brand of fertilizer is within reason. Persons who are unacquainted with the principles controlling the use of commercial fertilizers may, however, be led to believe that the price of the fertilizer is an indication of its value for the production of a given crop. As is well known to all students of the subject, there is no necessary relation between the commercial value of a fertilizer and the fitness of its formula for a given soil and crop. For these and other reasons, the publication of tables of commercial value has been strongly opposed by some manufacturers, and in certain states the custom has been discontinued. While granting that tables of commercial value are subject to misinterpretation, it is perhaps fair to say that such tables have been of most benefit, and, moreover, have been of great value to those who were most likely to misinterpret them.

It has been customary in most states to make analyses only of mixed fertilizers. Thus such raw materials as nitrate of soda, sulphate of ammonia, dried blood, bone meal, rock phosphate, tankage, muriate of potash, sulphate of potash, have not been brought under the operation of the law. If one wishes to purchase nitrate of soda, muriate of potash and tankage with the intention of mixing them according to a formula of his own, he may not find any protection in his state. However, these products can be obtained through reputable dealers who will willingly guarantee the contents. In case of doubt, the purchaser may secure an analysis by his state experiment station at a moderate cost.

The law requires that there shall be affixed to every package of fertilizer offered for sale a statement about as follows:

The minimum per centum of each of the following constituents which may be contained therein:

(a) Nitrogen.

(b) Soluble, available and total phosphoric acid, except in cases of undissolved bone, basic slag phosphate, wood ashes, unheated phosphate rock, garbage tankage and pulverized natural manures, when the minimum per centum of total phosphoric acid may be substituted. This latter applies only in those states where raw materials are subject to inspection.

(c) Potash soluble in distilled water.

It is possible to comply with the law and yet state the guarantee upon each bag of fertilizer in such a manner as to mislead the uninformed. It is not the purpose of this book to deal with such technical details, but if the purchaser of commercial fertilizers is not already well acquainted with fertilizer terms, he should secure an elementary textbook on the subject or write to his state experiment station for a bulletin discussing them.

FEEDING STUFF CONTROL

The law controlling the sale of stock foods is of more recent origin than the fertilizer control act and has not been so universally adopted up to the present time. The necessity for such a law arises from the growing use as stock foods of various by-products in the manufacture of liquors, starch, glucose, sugar, cottonseed and linseed oils and breakfast foods. Various mixtures, varying widely in chemical composition, especially in protein and crude fiber, were placed upon the market. In some instances mixtures were grossly adulterated with such things as oat hulls and ground corn cobs.

The adoption of this law by certain states has served to make other states the dumping ground for inferior stock foods, thus increasing the necessity for similar protection. The law does not apply to the ordinary grains produced by farmers or to the usual by-products of millers.

SEED CONTROL

From time immemorial it has been the universal custom of seedsmen to disclaim all responsibility for the purity and germinating power of their seeds. But as the importance of good seed--good in hereditary power, good in germination, good in its freedom from adulteration, good in its absence of noxious weed seed--has become better understood demand for some method of control has arisen. In at least one state there is a seed-control law modeled quite closely after the fertilizer-control law. However, the usual method of protection consists in purchasing by sample or the insistence of a guarantee, with a subsequent "analysis" of a sample of the purchased seed.

The germinating power and purity of seed can be determined cheaply by an expert within from five to twenty days, depending upon the species. The federal government has a division of seed control in its Department of Agriculture at Washington, D. C. Any person may send a sample of seed to this division and have its purity and germinating power determined, and in some of the states the experiment station will perform similar services without charge. Clover, alfalfa, grass and other small seeds should always be purchased subject to such inspection, unless the purchaser is prepared to make his own inspection, which a very little training makes possible.

NURSERY INSPECTION

There is no national law concerning the importation of insect-infested or diseased plant stock.

Several of the states have passed both state and interstate regulations concerning the sale of nursery stock. The insects usually legislated against are San Jose scale, gypsy moth and brown-tail moth, while the diseases usually interdicted are yellows, black knot, peach rosette, and pear blight.

The enforcement of the law is usually placed in charge of a person having special knowledge of economic insects and fungous diseases. In addition to these police regulations this officer may, by various means, attempt to bring into practice methods calculated to eradicate or, at least, lessen the severity of existing attacks.

Commerce in vinegar, dried fruits, insecticides and fungicides is also regulated in some states.

DAIRY, FOOD AND DRUG INSPECTION

An adequate discussion of the rise and development of the control in the sale of dairy and food products would require a chapter by itself, if not an entire volume. Suffice it to say here that the laws on this general subject have acquired an importance in many ways quite beyond that of any of the other control measures discussed in this chapter. In the extent of funds handled, the number of agents employed and the public interest incited, the office of dairy and food commissioner outranks any other control agency. In some states the office is an elective one, and the questions with which the office has to deal become a part of the state political campaign.

The importance of the inspection of dairy and food products grows out of the fact that not only is the consumer, hence all the world, interested, but the execution of these laws touch large commercial interests. Not only are meat packers, distillers and brewers deeply interested, but the wholesale and retail grocers and, more recently, the manufacturing and prescribing druggists, are vitally concerned.

Not many years ago the inspection of dairy products, particularly control of the traffic in oleomargarine, was the chief function of this office. To-day the enforcement of laws concerning pure foods, liquor and drugs is of much greater importance.

Interstate commerce in oleomargarine is now regulated through the enactment of an internal revenue law requiring a tax of ten cents a pound on colored oleomargarine and one-fourth of a cent a pound on uncolored oleomargarine and, further, by prescribing the character of package and method of marking all oleomargarine entering into interstate commerce. State agencies are charged with the duty of requiring the compliance of local dealers and restaurateurs with the general features of the federal law. Some states, however, prohibit entirely the sale of colored oleomargarine within the state.

PURITY IN DAIRY PRODUCTS

Attempts to define what is pure milk, cream, butter or cheese have been fraught with much difficulty. Thus, for example, legal definitions of pure milk have resulted in some cows giving illegal milk. In some instances the law has declared simply that whole milk is milk from which no cream has been removed; in others, the minimum amount of butter fat has been prescribed; in still others, the minimum amount of total solids containing a minimum proportion of butter fat has been made the basis of legal milk. In like manner full cream cheese has been defined as cheese made from whole milk or from milk from which only a given amount of cream has been removed, while in other instances the minimum amount of butter fat which full cream cheese may contain is prescribed. The wide variation in the amount of butter fat carried by cream has caused much jocular comment and some serious discussion as to what is cream.

While it is not feasible to indicate the laws for the several states, the ruling of the federal government as to what constitutes purity in dairy products under the national food and drug act may be accepted as a general guide. A circular giving the required information may be secured by addressing the Secretary of Agriculture, Washington, D. C.

LIVE STOCK SANITATION

The control of contagious diseases in domestic animals and the inspection of meat products have been the chief work of the Bureau of Animal Industry of the United States Department of Agriculture since its establishment.

The bureau inspects all imported live animals and under certain conditions will inspect live animals intended for exportation. It inspects all meat products intended for export. Its inspection of meats intended for interstate commerce is less rigid than that exported. Meats sold within the state in which they are slaughtered cannot be required by the federal government to undergo inspection. It thus happens that the people of the several states enjoy less protection in the consumption of meat than the foreign purchaser of American meats unless there is a state meat inspection law. However, it is becoming more and more the custom for the large packers to have all their products inspected without regard to their destination. The meats slaughtered in the locality in which they are consumed are the ones that receive the least supervision.

The federal government has been especially active and efficient in the prevention of interstate commerce in cattle suffering with Texas fever, and sheep attacked with scab and foot rot. Through the agency of the bureau dipping tanks have been provided in all the great live stock markets for the disinfection of cattle and sheep when needed.

Several of the states have laws controlling the importation of diseased animals from other states and the transfer of them within the state. The following are the diseases most commonly mentioned in the laws of the several states: Anthrax, black quarter, hog cholera, swine plague, rabies, glanders and tuberculosis. The law is generally enforced by a state veterinarian, whose acts are supervised either by a state live stock commission or the state secretary of agriculture or these two agencies acting conjointly.

Perhaps the disease which has required the greatest amount of attention in the several states is tuberculosis in milch cows. It is customary for this office to apply the tuberculin test, free of charge, under certain stipulations, to any herd upon the request of the owner and to supervise the slaughter and disposition of the reacting animals. In some states the owner is indemnified in part or in whole for his loss. The amount of indemnity as well as the general features of the law concerning the control of tuberculosis in domestic animals has been the subject of much controversy and cannot be said to have reached an altogether satisfactory solution in most states.

The young farmer should clearly understand that under no circumstances can he afford to have a tuberculous animal in h s herd. The contact of a diseased animal with other animals of the herd is certain to entail a greater loss than the destruction of the diseased animal. The farmer must in his own interest rear healthy animals whether or not it is necessary for the protection of the consumer.

FISH AND GAME LAWS

The motives underlying the enactment of laws concerning fish and game are varied. The controversies over these laws in the legislatures of the several states indicate that there is a belief, whatever may be the fact, that there are opposing interests; viz., those of the hunter or sportsman on the one hand, and those of the farmer or landowner on the other. The law of trespass has been one over which has raged much bitterness, both with regard to the form of the law to be enacted and concerning its subsequent enforcement. Sportsmen have usually held that a distinction existed between wild animals occupying private property and domestic animals. The landowner has urged that others should not trespass upon his property for the purpose of shooting wild animals, although his proprietary right in them was no greater.

In like manner, laws concerning the closed season, made to protect animals during the breeding period, are the subject of extended discussion and are being constantly changed; both because there is a difference of opinion concerning the habits of the different species and because the motive varies for maintaining the supply. Some animals are protected on account of their benefit, supposed or real, to agriculture. Other animals are protected because of their gaming qualities, even to the extent of sometimes injuring farm crops. The money spent by sportsmen in the pursuit of game is an element in the varied interests involved. Humane motives and a desire to prevent the further restriction of a not too varied fauna have helped, also, to save certain species from extinction. On the other hand, in some states commercial interests are involved, as where large quantities of birds are taken for their plumage.

Some attempts have been made to introduce foreign species, as the Japanese pheasant. It is, however, with fish that the most has been accomplished in replenishment. The federal government and several of the states have been active in regularly restocking, each season, certain streams with "fry" of edible and game fish.

Information concerning the open season can be obtained from the proper state officer. The fish and game laws are usually under the control of a commission with a secretary as the executive officer.

CHAPTER XXII

RURAL FORCES

The United States is a vast domain. Its material resources are enormous. Its fertile and easily tilled soil, its magnificent forests, its great stores of ore, coal, oil and gas; its fine water-power sites and its temperate and healthful climate have all contributed to the making of a prosperous and progressive nation. Without these natural resources the United States could not be what it is.

The waste of some of these resources is almost beyond belief. In mining, one-half the anthracite and one-third the soft coal is left in the ground in such a manner that it may never be economically recovered. A ton of coal will produce 1,400 pounds of coke, worth $1.50, and 20 pounds of sulphate of ammonia, worth 50 cents. If all the nitrogen in coal which is turned into coke in Pennsylvania were recovered, it would furnish enough of this element to supply the needs of every acre of tillable soil in that state. Only about 44% of the wood in the trees now harvested in the United States is incorporated into buildings, apparatus and furniture. The rest is wasted in the process of cutting, sawing and manufacturing into the finished products.

Facts like these have led the nation to realize that the conservation of our natural resources is an immediate and pressing problem. The United States has, however, a greater inheritance than these great and beneficent gifts of nature and a more fundamental problem than the preservation and efficient use of them. In a single sentence, the greatest inheritance of the American people is their Puritan ancestry. The word Puritan is here used to apply not only to the New England Pilgrims, but to all our early forefathers, whose traditions and practices have served to set this country apart from the other countries of the world. Because of the traditions which have been handed down to us, we are healthier-bodied and cleaner-minded men and women. We are more efficient, not merely in making money, but in everything that goes to make a full and well-rounded life.

It is well to realize the resources of other nations. The agricultural possibilities of France appear to the casual observer to compare favorably with any equal area in the United States. One may see farm land in Italy which has been cultivated for at least two thousand years which is evidently as fertile as any of the limestone valleys of the Atlantic States, the prairies of the Mississippi valley or the Palouse district of the Northwest. Russia has enormous areas of fertile soil. Careful observers report that in Manchuria there are great stretches of country, which today possess natural opportunities similar to those which the Mississippi valley offered one hundred years ago. The recent stories of the deposits of coal and mineral wealth in China are almost fabulous. Europe has rich mines, great forests and unrivaled water-power.

Some years ago a native of Argentina and a native of the United States were dining together. The Argentinian had served his government as consul to Canada. He related that he had recently written an official letter in which he had occasion to refer to the people of Canada and to those of this country. He explained that in alluding to the former he could say the Canadians, but the latter he could not call Americans, since his people were also Americans. After due consideration he referred to us as "the Yankees." "But," turning to his hearer, he said, with great emphasis, "I do not look upon the people of the United States as a nation, but as a new civilization." In other words, our nation is not simply one of fertile farms, enormous mines, great forests, unparalleled railroad systems, palatial stores, or wealthy cities, but he saw that we are a people of different economic, political, educational, social, moral and religious ideals.

There are in every rural neighborhood certain forces whose objects are to increase the educational advantages, the social opportunities and the moral aspirations of the people. This subject need not be discussed merely in the abstract. There are in every community concrete evidences of these forces. There is the rural church. There is the rural school. In many localities are to be found, also, buildings, for social and fraternal purposes, as grange halls, structures for holding fairs and picnics. These are tangible evidences that there are rural agencies at work in the community whose chief purpose is to increase the educational advantages, the social opportunities and the moral aspirations of the people.

How are these existing rural forces to be made more effective? If co-operation in financial affairs is essential under modern conditions, it is more needed in social matters. Such co-operation does not imply that these separate forces shall be fused into a single one. Each of them has its particular and peculiar work to do, but each should work in harmony and not in the spirit of antagonism with the others.

There should be formed in each locality a committee for which the following name is proposed: The Community Committee of Rural Forces. Emphasis should be placed upon the word "community." Like all moral movements, progress must come from within, and not from without. The movement must be adapted to its environment. Like the plants that grow there, it must be indigenous to the soil.

This committee should be composed of representatives of the churches, the schools, farmers' clubs, granges, fair associations, farmers' institutes; and other organizations which are striving to increase the educational advantages, the social opportunities and the moral aspirations of the people.

Oftentimes the object of these rural forces is confused with efforts to increase the financial prosperity of the farmer. It goes without saying that the maintenance of the fertility of the soil is essential to the food supply of the nation. The problems of the economic production of plants and animals are of great importance to the prosperity of the farmer. The idea, however, that the proper solution of these economic problems is to be the means of solving the educational, social and religious problems is simply putting the cart before the horse. Economic questions can only be satisfactorily adjusted through the application of intelligence and right ideas.

Let it be supposed that when a young man decides to pay attention to a young woman that instead of meeting her at the church door, or it may be at the railway station, it is considered better form for him to get permission of the mother to call upon the young woman in her own home. This is the most fundamental question in every neighborhood. What has it to do with the price of wheat?

This illustration has been used to emphasize two points. First, there are many problems in every community that are in no way related to the material prosperity of the neighborhood. Second, there is, at present, no single force in the community with sufficient influence to cope properly with many of these problems.

A young college graduate who is now managing eight hundred acres of land recently wrote: "I firmly believe that one of the best opportunities to be of help to a rural community lies in the work that is to be done for the improvement of social conditions--to help make what little leisure there is clean and refreshing." Hence on return from college this young man has found time to play football and baseball with local teams and to help whenever opportunity offered at dances, musicales and similar entertainments. Games and other forms of recreation may be clean and wholesome, or they may be quite the reverse. It would be the duty of the community committee to see that dances occurred under proper environment--not next an open saloon--and that the young women were properly chaperoned.

In many communities the boys and girls are almost wholly dependent upon the neighboring towns for their amusement. This condition may or may not be desirable. If the town and country are virtually one community, there is every reason why the boys and girls from the farms should find recreation and social intercourse with the boys and girls of the village. It is a relationship that should be fostered wherever possible. When, however, the town and the country are separate communities, which prevent the ordinary social relationships, it is usually unfortunate when the young people of the one community are dependent upon the other community for their amusements.

A deeply earnest man recently said: "I was born and raised upon the farm. I never knew a dull day in my life. I went fishing. I went hunting and----"

"Stop right there," said the listener. "There is not the same opportunity today for a boy to go hunting that there was when you were a boy."

"That is true."

"Our ideas about such things have changed, also."

"Yes," he replied, humbly enough, for he was a man of fine fiber.

"I propose a substitute," said the listener. "There is much more pleasure and recreation to be obtained from photographing animals than from killing them. What is needed in every rural community is a camera club."

When a boy wishes to go hunting, he merely has to buckle on his ammunition pouch, shoulder his gun and he is ready. A camera club, however, requires a social organization and a social center. The community committee would thus be required to decide whether the facilities for developing and printing pictures may best be located at the church, the schoolhouse, the grange hall or elsewhere.

A little reflection will show how many possibilities such a club might have on its social, moral and educational side. The suggestion has been made here, however, only as an illustration of the problems which arise when a rural community is organized for social welfare. The organization of a book club, or a magazine club in a rural community presents precisely the same problems. Some method must be devised for exchanging the books or magazines. Whether they are exchanged at the church, the grange hall or through the school children will depend upon local conditions requiring a community committee to decide.

This community committee will do something more than reach immediate results. It may project its influence far into the future. Not all of life is comprised in a porcelain bathtub and nickel adornments. Nevertheless modern methods of heating and plumbing are desirable in the country as well as in the city. In Indiana there is a one-room school building. In the basement there has been placed a furnace and a gasoline engine. The engine is used not only to teach the boys how to run a gasoline engine, but it makes possible a modern system of plumbing.

It is well known that many of the states within the past decade have voted to abolish or very materially restrict the sale of alcoholic beverages. No great temperance orators have roused the people as was the case thirty years or more ago. Why, then, has such progress been made in recent years? In large part because twenty-five years ago, the teaching of physiology was introduced into the public schools, which taught the evil effects of alcohol to the human system. During the past decade young men who studied these physiologies have been voting.

What has the teaching of physiology to do with the one-room schoolhouse in Indiana with its modern system of plumbing? The girls between the ages of six and fourteen are now becoming accustomed to modern systems of plumbing. When they grow older and marry they will find some way to introduce similar conveniences into their homes without regard to the price of wheat. A wise community committee will find many ways to influence future generations. Such a committee would be a priceless heritage to any community.

The natural resources of the United States are necessary to the prosperity of the people. The preservation and economic use of these resources are of vast importance. The natural resources of the world were, however, as great five thousand years ago as they are today. The soil was no less fertile then than now. The difference between the prosperity of the human race at these two periods is caused by a difference in human motive and efficiency. It is the result of ideals and knowledge. Sit at the banquet table with men who are the real powers in shaping the affairs of the world. The chances are that the champagne remains untouched. These men are not in the habit of partaking of midnight suppers. They must keep themselves fit for the next day's work. They have the approval and loyalty of their wives because they deserve it. In other words, the men who do the world's work are not drunkards. They are not gluttons. They are not libertines. They are efficient because they have healthy bodies and clean minds. It is this efficiency which the critic from Argentina saw when he said, "I do not look upon the people of the United States as a nation, but as a new civilization."

###

www.ingramcontent.com/pod-product-compliance
Lightning Source LLC
Chambersburg PA
CBHW070323190526
45169CB00005B/1726